Hiding Behind the Fog:
Stop Ignoring Cognitive Impairment in Autoimmune Disease

by Julie Owen Morris, Ph.D.

© Julie Owen Morris, B.S., M.Ed., Ph.D., CCC-SLP, NBC-HWC

Copyright © 2025 by Julie Owen Morris, B.S., M.Ed., Ph.D., CCC-SLP, NBC-HWC
All rights reserved.

No part of this publication may be reproduced, distributed, or transmitted in any form or by any means — including photocopying, recording, or other electronic or mechanical methods — without the prior written permission of the publisher, except in the case of brief quotations embodied in critical reviews and certain other noncommercial uses permitted by copyright law.

For permission requests, contact the publisher at: "Julie Owen Morris"
julie@functionalautoimmunity.com

Title: *Hiding Behind the Fog: Stop Ignoring Cognitive Impairment in Autoimmune Disease*
Author: Julie Owen Morris, B.S., M.Ed., Ph.D., CCC-SLP, NBC-HWC

ISBN 979-8-218-69197-4

Printed in the United States of America
First Edition, 2025

This book is intended for informational and educational purposes only and is not a substitute for medical advice, diagnosis, or treatment. Readers should consult their healthcare provider before making decisions regarding their health.

Dedication

It is my privilege to dedicate this book to the many–too many–autoimmune patients, who suffer daily. I fear the number is only growing with long-COVID autoimmunity issues, and this makes me both sad and angry. Adding decreased cognition to pain, fatigue, and the endless other symptoms of autoimmune disease seems crushingly unfair. As someone who has been there, and is there, I want you to find hope in these pages. Although it's not moving nearly fast enough, research is being done, treatments are being developed, and help is already available. What's even more encouraging is that information in these pages will enable you to take control of certain factors that will have a real, positive benefit on your cognition, as well as your general health. Allow your family and friends to share the burden, but realize that you are the CEO of your own health. The resources in this book can help you take control of your brain and your health and work your way back to yourself.

*** Disclaimer: This book is not meant to diagnose, treat, or mitigate any disease, but is purely educational. Please consult your doctor before changing your diet, adding supplements, or modifying other lifestyle factors.

Acknowledgments

Without the constant strength and hope graciously given to me by the Lord Jesus Christ, I could not have endured multiple daily symptoms of lupus and other autoimmune diseases—the worst of which was mild cognitive impairment. God is my strength. I am grateful He continues to provide what I need daily. Like most autoimmune diseases, lupus has its better days, bad days, and horrible days. May God, through science, provide a cure so his children with autoimmune disease may again experience every day as a good day.

I deeply appreciate the support of my family: my extraordinary best friend and husband Chad; my generous, intelligent, and beautiful daughters Addison and Makenna; my loving, supportive parents Gayle and Bill, and my terrific brother Grant, as well as my wonderful in-laws Troy and Linda, and sisters-in-law Shelley & Lisa. Thank you all for tolerating my endless texts, emails, and lectures on brain health, and don't expect them to stop anytime soon! Thanks, also, to my wonderful cousin, Brandi, for her early editing, my close friend and kindred spirit, Gala, for editing several chapters, and my new friend, Sophie, for her suggestions. Thank you to Fatima for taking this editing project on, and doing a fabulous job with it!

Thank you to my team of doctors, who have worked tirelessly to make my life better. A special thanks to Dr. Craig Carson, my rheumatologist for 18 years, who wrote the first forward. I believe Dr. Carson saved my life at least twice, and he is the most kind, compassionate doctor I have ever met. I am also very grateful to Dr. Todd LePine, my functional medicine doctor for the past 6 years, who wrote the second forward. With his research-backed intelligence, he was able to solve the puzzle of mild cognitive impairment for me personally, adding nutrition, supplements, and innovative treatments to improve my cognition and my autoimmune symptoms dramatically. In addition, other doctors keep me functioning week to week—my chiropractor, my neurologist, and a host of other specialists. Thank you for your care and your willingness to take on a "difficult case" with six autoimmune diseases.

Thank you to my former professors, Sarah, Susan, & Judy, who taught me how to research and modeled an obsessive pursuit of knowledge. Thank you to all the researchers who continue to work tirelessly to unravel autoimmune diseases and cognitive impairments. Your work is what this book is based on, and the progression of your work gives patients hope for a better tomorrow.

Thank you to all the lupus and other autoimmune patients who have shared their stories of struggle with me. Remember that your lives are **not** over with these diagnoses. They are different than what you had planned, but they can still be meaningful and full of life and love. When you can't think clearly through the fog, ask for help and remind yourselves that you are still you. You **can** improve, and you will take gradual steps to become the best version of you again. Never, ever, ever give up!

Table of Contents

Foreword (by Conventional Rheumatologist) V
Foreword (by Functional Medicine Doctor) VIII
Introduction: Don't Take My Brain, Too! 1

Part I: Understanding Lupus and Cognitive Impairment

Chapter 1: What Is Lupus? 9
Chapter 2: What Are The Treatment Options For Lupus? 17
Chapter 3: What Causes Lupus? 26
Chapter 4: What Is Cognitive Impairment? 40
Chapter 5: How Does Cognitive Impairment Appear In Lupus? 52

Part II: Investigating the Causes

Chapter 6: Where Is The Problem In The Brain? 61
Chapter 7: What Causes These Issues? 68
Chapter 8: How Do You Identify Brain Fog In Lupus? 84
Chapter 9: What About Domains, Demographics, & Quality Of Life? 103
Chapter 10: What Are The Cognitive Treatment Options? 109

Part III: Treatment And Lifestyle Alternatives

Chapter 11: Will Medications, Biomarkers, or Supplements Help? 119
Chapter 12: What Else Can We Do To Improve? 135
Chapter 13: How About Quality Of Life & Nutrition? 143
Chapter 14: Will It Progress & Can We Prevent It? 148
Chapter 15: What Can You Do At Home? 153
Chapter 16: What Are The Types Of Cognitive Rehabilitation? 175
Chapter 17: Do People With Other Autoimmune Diseases Get Cognitive Impairment? 188

Conclusion: What Can We Do To Help? 198
Epilogue 205
Glossary 210
References 216

Foreword

From a conventional rheumatologist:

Regarding Cognitive Impairment in Lupus
As a conventionally-trained clinical rheumatologist who has been practicing for 30 years, I have certainly recognized cognitive impairment as a common symptom of Systemic Lupus Erythematosus (SLE). It infrequently progresses to what would be considered dementia, however. In my efforts to reassure patients that it is a common symptom and not as bad as Alzheimer's, I hope I have not discounted the very real effect it has had on their daily lives. As mentioned in this book, cognitive impairment happens so frequently in lupus that it can affect up to 90% of patients at some point in their disease process.

Hiding behind the Fog: Stop Ignoring Cognitive Impairment in Autoimmune Disease by Dr. Julie Owen Morris is an insightful and engaging mix of personal anecdotes and the latest scientific information. This is a thoughtful review of the effects of MCI (mild cognitive impairment) on cognitive abilities needed in daily life, and it has helped me more fully understand the impact of this symptom on personal lives. The observation that mild cognitive impairment in lupus is evanescent, rather than progressive, fits with my experience as a clinician.

I particularly like the chapter in this book on etiology, which fits with my own clinical research. There is definitely an inflammatory autoimmune component which results in small vessel disease, but most often it does not result in permanent structural brain damage.

I also thought the discussion of the connection between leaky gut and food intolerances to brain function was at once thoughtful, scientific, and relevant. The brain is an obligate aerobe mechanism, meaning that it must utilize glucose selectively as an energy source, while the presence of sufficient oxygen is required to extract that energy through cellular metabolism. The fact that solid nutrition is required to feed the brain properly makes intuitive sense to me.

We truly are missing an effective treatment for MCI in autoimmune disease at this time. The current medication options available to address cognitive issues are admittedly ineffective. To address cognitive impairment in lupus in my own practice, I have tried to follow a treatment algorithm that includes the following: education, sun avoidance, diet, exercise, and, as a last resort, medications, due to poor efficacy. I have observed disappointing results with steroids, but have seen encouraging responses to Plaquenil, Benlysta, and even anticoagulants for certain patients. It has been difficult for me to find and recommend resources for cognitive exercises. Likewise, I have not routinely recommended speech-language pathology treatment or cognitive behavioral treatment for three main reasons: it is not pharmacologic; it is not readily available; and it is not reimbursed appropriately through insurance. We definitely need more available and effective treatment options in the clinic.

While I wish we had the perfect tests and treatments for cognitive impairment in SLE, we do not. Dr. Morris's book helps us understand the process in a very scientific way, and then leads us to a very personal and practical self-directed path to wellness, no matter the specifics of the condition. In line with these recommendations, I would suggest that patients not be passive in their treatment, and learn to take control of as many factors as possible.

I admire Dr. Morris as a professional who has dealt with chronic disease for many years, and yet maintained a healthy balance of work, family life, and personal care. I highly recommend her book for healthcare professionals who sincerely want to help their patients with this under-recognized symptom, as this book is thoughtful, organized, and scientifically-based. I would also recommend this book for patients, as it is accessible and directly applicable to lupus patients and their struggle with cognitive impairment. I have many patients who are distressed by cognitive impairment, and they do not know where to turn for answers. This book will give them a starting point.

Craig Carson, MD
Board Certified in Rheumatology & Internal Medicine
Founder of Oklahoma Arthritis Center

Foreword

From a functional medicine doctor:

"Brain Fog" is not an accepted diagnosis in mainstream medicine, but nonetheless it is a real phenomenon. The closest to giving a diagnosis for what patients call brain fog is the term cognitive impairment. Julie Morris, as both a PhD and a patient with autoimmune disease, knows first-hand what it's like when a sharp mind refuses to fire on all four cylinders. Autoimmune diseases are not well understood, but what we do know is that there are many types of autoimmune diseases with myriad presentations and that they occur more often in women and can be associated with stealth or low-grade infections from bacteria, viruses or atypical bacteria in genetically susceptible individuals. Not only does Julie Morris have Lupus, but she also has five other autoimmune conditions, the most I have ever seen in an individual. In her book, she outlines the approach she took to help create an environment for self-healing and calming down an overactive immune system.

We took a team approach using a variety of integrative approaches to quiet down her immune system. This included dietary approaches, managing stress, prioritizing sleep, healthy exercise, and directed supplements based on advanced laboratory testing. In her case, we were able to help a lot with initial Functional Medicine approaches, but because of her multiple autoimmune issues, what helped to move the needle the most with her was combination peptide therapy, which included thymosin Alpha and thymosin Beta 4. When there is inflammation in the body, it can manifest in a myriad of ways, including joint pain, muscle pain, sleep disturbances, and "brain fog" from inflammation in the brain. In layman's terms, this is the "brain on fire" and, in more severe cases, it can manifest as lupus encephalitis. Thankfully, using a multidimensional approach put out the fire and put an end to Julie's brain fog. It is my sincerest hope that many patients will find hope and help by learning from the story that Julie has to tell.

Todd R. LePine, MD
Functional Medicine Physician
Ultrawellness Center

Hiding Behind the Fog:
Stop Ignoring Cognitive Impairment in Autoimmune Disease

by Julie Owen Morris, Ph.D.

Introduction

Don't Take My Brain, Too!

I have been living with lupus symptoms for at least 20 years but was diagnosed 18 years ago. About 8 years ago, I began experiencing "Brain fog" or what I now know is mild cognitive impairment (MCI). I experienced my first symptom of MCI, word-finding difficulty, while I was lecturing in a university classroom. This wasn't a one-time occurrence but quickly became a frustrating pattern.

Up to that point, lupus had affected my joints, my kidneys, my lungs, my stomach, my colon, and my nerves. My overall quality of life was affected daily, through pain, fever, & fatigue. I was determined that it wasn't going to take my brain, too!

It took me about a year before I was brave enough to mention my concerns to my husband and my rheumatologist. My husband assured me that I was still much smarter than him. My doctor told me that brain fog is a common symptom of lupus. Honestly, that didn't help much. I didn't like it—in fact, I hated it. I knew my brain had changed and was less efficient than it used to be. Despite the daily fatigue, pain, and fever of lupus, MCI quickly became my most hated symptom. I had always been a quick thinker with a good vocabulary, but now I was struggling with finding words to express my thoughts and noticing that my receptive language processing was slowing down. I had completed 12 years of higher education, earning a bachelors, a masters, and a doctoral degree. I was not used to struggling cognitively.

Hiding Behind the Fog

As my brain continued to get worse, I was determined to find out more information. Online lupus support group discussions often talked about "brain fog" with lupus, but this symptom was often described in a mostly joking manner. In some of the posts, I could sense frustration, but most just blew it off as an annoying, yet light-hearted symptom. I asked myself, were these people really okay with this, or were they just too embarrassed to be honest about the extent of the problem? I could certainly understand the latter. An internet search revealed some acknowledgement of brain fog as a possible symptom, but most sites dismissed it as one of the lesser or insignificant symptoms. None of the sites presented any hope for treatment, but, instead, increased my anxiety. As stated by Tamilou, et al., "while a huge amount of information is available to patients on the web, this information is of very low quality, often counterproductive and anxiety generating."[1]

When I began having these cognitive symptoms, I had been a speech-language pathologist for over 20 years. Many people do not know that speech-language pathologists (SLPs) do not only work on speech and language skills (and swallowing and voice and fluency and accent reduction), but also on cognitive (thinking) skills. I had worked with many adult patients over the years and billed for cognitive assessment and treatment. These patients had diagnoses ranging from stroke to traumatic brain injury and dementia. However, I had never had a lupus patient seeking treatment for cognition. Yet, here I was, now a patient myself, and I quickly realized that what I had was not a benign, annoying "brain fog" but what seemed to me to be a measurable diagnosis of MCI. I took two online tests, which confirmed I was having cognitive impairment. The cognitive deficits were negatively affecting my daily life and I was beyond frustrated.

In addition to the word finding and slower processing symptoms, I began to notice other cognitive problems that I was having, such as difficulty with sustained attention, working memory, and math calculations. I went to my rheumatologist and requested an MRI and a more aggressive treatment for my lupus.

Introduction

The MRI was normal. The normal finding was both a relief and a frustration for me because I knew my brain was not functioning normally. I also knew from my PhD courses in neuroscience that an MRI would not necessarily pick up minor changes and that MRI machines differ significantly by resolution quality. About six months after receiving IV Benlysta infusions, I began to see an improvement, although no resolution, to my cognitive difficulties. After two years of this same monthly treatment, the cognitive difficulties began to creep back in more often, but were inconsistent in severity. I couldn't figure out what was going on and why, and I began to speculate that it might have something to do with overall systemic inflammation, as my C-reactive protein (CRP) levels had risen to dangerously high levels again.

Not being satisfied with the information found on lupus websites, I decided to put my PhD research skills to use and began to delve into published research studies. Looking back, I was seeking verification for what I was experiencing, explanations for the specific cognitive deficits I had, and, most of all, hope for proven treatments to address the deficits. What I soon discovered is that there wasn't (and isn't) near enough research out there (particularly in the area of treatment), but there *have* been some recent efforts to investigate cognitive impairment in autoimmune patients. I've learned a tremendous amount over the last several years, reading and summarizing articles and books on autoimmune disease, mild cognitive impairment, Alzheimer's, dementia, nutrition, and brain health.

The secret to surviving my autoimmune journey so far has been a combination of conventional and functional medicine. Some people insist on going one way or another, but I've needed both. Conventional medicine saved my life when I was really sick and the lupus was affecting my kidneys (with proteinuria or protein in the urine) through Methotrexate. The medication had horrible side effects, but it worked. Conventional medicine also offered me help through Benlysta infusions when I was having neurological symptoms, like peripheral neuropathy and the beginning of autonomic neuropathy. I believe the Benlysta infusions stopped the progression of the disease. However, it wasn't enough to consistently address the brain fog or MCI. To fully address my cognitive difficulties, I needed functional medicine. A few years ago, I saw a functional medicine doctor, had a thorough workup, and, thanks to a specific diet, supplementation, and a healthy lifestyle, I no longer suffer from cognitive issues. My worst symptom is gone! Happily, along the way, fatigue, joint stiffness, and gastrointestinal symptoms have also significantly improved through functional medicine. In addition, I've been able to get off nine conventional medications. For me, then, a combination of conventional and functional medicine has been the most effective means to achieve a healthier, more fulfilling life, and a healthy brain.

Why Write This Book?
I'm writing this book, first and foremost, for people with lupus and other autoimmune diseases. My sincere hope is to help fellow autoimmune sufferers to understand that they're not crazy; that what they are struggling with is common and real; and that it matters. If you are suffering from cognitive dysfunction, it matters to you, to your life, to your family, and it matters to me. I hope to help you understand what is going on in your body and what researchers are discovering about cognitive impairment in lupus and other autoimmune diseases.

Second, and most importantly, I want to instill in you some hope, by helping you discover that there are things you can do to improve your daily life, and by understanding that there is current progress being made in research and treatment. The first part of the book is pretty technical (some might even say boring), as I summarize and describe the current literature. Since I am not an immunologist, I have done my best to summarize the information I gleaned from the literature. If you have a background in biochemistry or immunology, please look up the listed studies for yourself, in order to obtain more details. Alternatively, if science is

Introduction

not your thing, feel free to skip to the treatment sections (chapter ten on) whenever you want, so that you can explore what is available for you. I want you to know that there are ways you can help yourself and there is reason to hope for better treatment in the future. Lupus warriors are tough and together we can fight cognitive impairment! Will you join me?

During my research, I discovered that cognitive impairment is a symptom in other autoimmune diseases, as well. More recently, some scientists are drawing a connection between COVID-19 and autoimmune disease, particularly in long-COVID (see Knight, et al., 2021).[2] We have all heard on the news about the "brain fog" issues in the long-COVID population. Although I began writing this book before the pandemic, people have recently become more aware of the general concept of "brain fog," due to this common symptom of COVID (see Apple, et al., 2022), particularly amid the long-haul syndrome accompanying the disease.[3] I hope that this book will be beneficial to anyone who is seeking answers for their "brain fog." If you have a friend or family member who is suffering from this overlooked symptom, I hope that you will share both the information and the hope with them.

My third goal in writing this book is to open the eyes of speech-language pathologists (SLPs), neuropsychologists, doctors, and other professionals to the significant need for cognitive treatment in people with lupus. Although the Centers for Disease Control (CDC) does not have recent data for the prevalence of lupus, citing expense and difficulty tracking diagnoses, older national estimates suggest a national prevalence of 322,000 people with definite or probable systemic lupus erythematosus (SLE) (see https://www.cdc.gov/lupus/facts/detailed.html#prevalence).

Another group of researchers discussed SLE as a severe autoimmune disease, and estimated its worldwide prevalence at .02-.24%.[4]

Many people with lupus and other autoimmune diseases deal with cognitive impairment daily, and most face it alone. SLPs need to be aware of the challenges of these patients and be able to provide appropriate treatment and support for autoimmune patients with cognitive impairment. I hope this book will introduce you to cognitive deficits in lupus and spark your interest in learning how to help these patients. I also hope it will inspire researchers in the field to investigate various types of cognitive treatment specifically for autoimmune patients. Colleagues, we need evidence-based, effective treatment options for this population.

This book is based in research, as I have read, cited, and summarized information from numerous studies and books over the last seven years. This book is timely, as neuropsychiatric systemic lupus erythematosus, including cognitive dysfunction, "is an emerging frontier in lupus care."[5] Throughout the book, I have also peppered personal information from my own story. I hope that this is not too distracting for you, as I switch between professional and personal voices. Since my main purpose for writing this book is to help provide information and comfort to lupus patients who are suffering, I wanted to share with them that they are not alone in this terrible struggle.

Introduction

Never intending to share this with the world, in October of 2014, I wrote the following in my journal:

Journal Entry (October, 2014)

"Brain fog is the absolute worst symptom of lupus—worse than fever, pain, fatigue, or dizziness. It is THE WORST! It's like I don't recognize my own brain anymore. If feels fuzzy and full. It's like you are trying to draw a conclusion or solve a problem or find a specific word, yet you can't get through your brain to get to it. You are trudging through something thick and disturbing. You know you could normally do this quickly and easily, but now it is so painstakingly slow, unsure, and unreliable.

What does this mean physiologically and chemically? What is really going on in there? When pushed, my neurologist agreed that it was probably brain swelling, yet gave me no idea what to do beyond "get your lupus under control." Is something slowing my synaptic connections? Are my neurons dying? Is this progressive?

Sometimes I even feel a physical pressure or a buzzing in my head. It is strange and unsettling. Frightening because I can't stop it. No idea how long it will last.

What if this is doing permanent damage to my brain? What if the damage accumulates and causes dementia? What if I can't work anymore? What if I can't take care of my family?

I feel such shame and embarrassment in this, yet I know it's not my fault. What can I do to help myself?"

My answer to myself several years later? Search for answers; find help; and hope. Then, publish a book to help others who are equally frightened of this awful "brain fog."

Part I:

Understanding Lupus and Cognitive Impairment

Chapter 1

What is Lupus?

Lupus, specifically systemic lupus erythematosus (SLE), is a complex and often misunderstood autoimmune disease that can affect nearly every organ system in the body. At its core, lupus results from a malfunction in the immune system, where instead of protecting the body, it turns against it, causing widespread inflammation and tissue damage. While the disease is known for its unpredictability, relapsing, and remitting in nature, its manifestations are both diverse and serious, ranging from joint pain and skin rashes to kidney disease and neurological complications.

Diagnosing lupus remains a major challenge due to the broad spectrum of symptoms and overlapping signs with other conditions. As research continues to evolve, so does our understanding of its causes, diagnostic criteria, and the factors influencing its progression. This chapter explores what lupus is, how it affects the body, how it is diagnosed, and the realities of living with such a life-altering condition.

The Basics of Lupus

Systemic lupus erythematosus (SLE) is a horrible multiple-organ, systemic disease. The immune system is overactive in SLE and can negatively affect any system or organ.[1] As we know, all autoimmune diseases stem from a breakdown in tolerance towards self.[2] In lupus, the immune system becomes overactive and causes the body to attack itself. This can damage the skin, joints, kidneys, heart, lungs, central, autonomic, and peripheral nervous systems, and more. Moreover, it typically has a relapsing-remitting pattern and is entirely unpredictable.[3] Affecting body systems, SLE is caused by autoantibody production and complement-fixing immune complex deposition, which damages tissues.[4] Stated another way, "systemic lupus erythematosus is an autoimmune disease characterized by antibodies that bind target autoantigens in multiple organs in the body. In peripheral organs, immune complexes engage the complement cascade, recruiting blood-borne inflammatory cells and initiating tissue inflammation."[5] A complement deficiency has been found to predispose an individual to SLE.[6]

Based on this understanding, Yuen and Cunningham have found that SLE is "characterized by production of pathogenic autoantibodies and dysregulated immune responses by B-cells, T-cells, dendritic cells, and other immune cells, resulting in numerous clinical and serological manifestations."[7] The earliest autoantibodies have been found to attack RNA binding autoantigen Ro60, but researchers don't know why.[8] Autoantibodies may "form pathogenic immune complexes" as a result of certain T cells and B cells that stimulate production of these autoantibodies.

This autoantibody production may result from a deficit in the immune system of lupus patients, in which nucleosomes (unit of a chromosome/length of DNA) from apoptotic (dead) cells become major immunogens.[9] It is thought that a deficit occurs early in the life of the lupus patient, in which the immune system is unable to clear dead cells, and these cells become immunogenic. This lack of clearance happens because scavenging molecules, whose job is to clear apoptotic cells, are deficient.[10] Complement components are also deficient, which are also supposed to help clear dead cells through phagocytosis (immune phagocyte cells swallow/ingest other cells).

Basically, in lupus patients, the immune system often creates autoantibodies, which are antibodies that mistakenly attack the body's own tissues. Again, one reason this happens is because the immune system doesn't properly clean up dead cells. When cells die, in a normal process called apoptosis, their remains should be quickly cleared away. But in people with lupus, this cleanup process doesn't work well, even from an early age. The body lacks enough "scavenger" molecules and important immune components called complement proteins that are supposed to help get rid of these dead cells. Because the dead cells stick around too long, parts of them, especially pieces of DNA called nucleosomes, end up triggering the immune system to treat them like dangerous invaders. This confusion can cause the body to attack itself, leading to the chronic inflammation and symptoms seen in lupus.

Diagnosis of Lupus

The daunting task of diagnosis in such a complex disease has gone through several revisions. The American College of Rheumatology (ACR) revised diagnostic criteria in 1997 such that four of eleven ACR criteria must be present either serially or simultaneously for a diagnosis of SLE.[11] These eleven criteria are malar rash, discoid rash, photosensitivity, oral ulcers, arthritis, serositis, renal disorder, neurologic disorder, hematologic disorder, immunologic disorder, and antinuclear antibody.[4]

The ACR first proposed this criterion in 1971 and underwent two changes in 1982 and 1997. Then, in 2012, the Systemic Lupus International Collaborating Clinics (SLICC) proposed new SLICC criteria for SLE due to new research knowledge on autoantibodies and low complement.[12] As per SLICC criteria, at least one clinical criterion and one immunologic criterion are to be present in a patient, with a total of four criteria.[13] Alternatively, a diagnosis of lupus nephritis as the sole symptom, along with ANA or anti-dsDNA antibodies, is another option. The 17 SLICC criteria are less specific and more sensitive than the 1997 ACR criteria, but "both recognize the principles that SLE is characterized by the presence of autoantibodies and can cause clinical effects in multiple different organs or tissues."[14] SLICC criteria took over a decade to develop, and data from 1300 patients were pulled together.[15]

In addition, classification criteria for SLE jointly supported by the ACR and the EULAR (European League Against Rheumatism) were adopted in 2019.[16] These diagnostic criteria include a positive ANA (antinuclear antibody test) and weighted criteria in 7 clinical and 3 immunological categories. This complex method of diagnosis claims to have good sensitivity (how often it generates a correct positive result) and specificity (how often it generates a correct negative result) and is beneficial for the continuity of diagnosis since America and Europe worked together on these criteria.[16]

Despite these advancements, diagnosis of lupus is a complex process. However, if verified, recent research identifying oral microbiota dysbiosis (reduced microbial diversity and inappropriate balance) in lupus patients might mean an easier diagnostic process.[17] There are so many symptoms and so much variability among individuals with lupus that researchers have recently begun to divide symptoms into Type 1 (rashes, nephritis, arthritis) or Type 2 (cognitive dysfunction, mood changes, myalgia, fatigue). However, a recent study concluded that there are also two different patterns of Type 2 patients (intermittent or persistent).[18]

Unfortunately, SLE is incurable. The population most often affected by it are young biological females of child-bearing age.[19] In fact, the peak age of onset is thirty years.[20] A study has defined SLE as a "chronic multisystem inflammatory autoimmune disease with a waxing and waning course and a broad spectrum of clinical presentations."[21] The "waxing and waning course" refers to periods of both flare-ups and remission of symptoms, although, in my experience, it is more like increased disease and decreased disease. I have never experienced a complete remission of symptoms, and I wonder how many lupus patients have a complete resolution of symptoms, as well as normal lab values. From talking with other lupus patients, I know there is a broad spectrum of severity of clinical manifestations, ranging from very mild to extremely severe. Furthermore, studies have revealed that lupus is "persistent and recurrent."[22] This aligns better with my experience, as it persists in my body in some form always but has flared up into a severe form multiple times for extended periods.

It is thought that lupus symptoms result from the presence of high titers of autoantibodies. These autoantibodies accumulate in the tissues and then form immune complexes, activating the production of immune system cells that perpetuate a feedback loop, resulting in organ damage (often of vital organs such as the brain, heart, joints, skin, and kidneys).[23] Inflammation is a large part of the process. Some of the most common manifestations (of which I've had all) are arthritis, serositis, fever, mucocutaneous lesions, renal involvement, and hematological disorders.[3]

The Tragedy of Lupus

A recent article gave a depressing definition of SLE as "a potentially fatal autoimmune disease that is often accompanied by brain atrophy and diverse neuropsychiatric manifestations of unknown origin."[24] The mortality rate for lupus patients is still 3-4 times higher than that of the general population.[25] This is primarily because in severe presentations of the disease, multiple organ systems (CNS/kidneys) are affected, and these are what cause the increased mortality and morbidity.[13, 26, 27]

The fact that increased morbidity and mortality are still facts for SLE is inexcusable. According to recent studies, it may be a result of delayed onset of diagnosis and a lack of consistent and coordinated care.[28] It is one of the leading causes of death in young women, with lupus often contributing to cardiovascular disease and severe infections.[29] Lupus has been described as "one of the most disabling autoimmune pathologies known to have an effect on the central nervous system secondary to the systemic disease is systemic lupus erythematosus."[30]

Despite the variability in outcomes of lupus patients, the overall risk of death continues to be increased for these patients. Ten-year survival rates increased from 1950 to the mid-90s but then plateaued.[25] This is probably due to the lack of new treatments.

Epidemiology of Lupus

Systemic lupus (SLE) is somewhat rare and varies by gender and ethnic group, but it has been known to occur in people of all ages and races.[27] First, let's look at the prevalence of autoimmune disease in general, with a recent study indicating that over 15 million people in the United States (4.6% of the U.S. population) had been diagnosed with at least one autoimmune disease from 2011 to 2022, with 34% of those having more than one, and females were nearly twice as likely as males to have an autoimmune condition.[31]

Focusing specifically on lupus, a recent meta-analysis estimated specific SLE/lupus incidence in the United States of America at 5.1 per 100,000 people.[32] The global incidence ranges from 1.5 to 11 per 100,000 people.[25]

Moreover, it is well established that more females than males have lupus at a ratio of 9:1. However, SLE in males is rising.[33] Lupus tends to strike females ages 16-55 years.[3] However, it can develop at any age. In a recent study, the median age at diagnosis was 31.2 years.[34] Lupus is highest among black females in the United States, followed by American Indian/Alaska Native females.[32] In a population analysis in Michigan, SLE prevalence was 1 in 537 for black females.[35]

Lupus varies, then, by ethnic groups in North America, affecting non-Caucasian populations more often: American Indians, Hispanics, and Asian people tend to have more neuropsychiatric issues with SLE, although white patients may have more overall neurological damage.[36] Non-European ancestry and lower socioeconomic status are both associated with poorer outcomes in lupus, and patients of African ancestry, Hispanics, and Asians have a significantly higher rate of death from the disease.[27] SLE patients with Asian heritage have also been found to display more severe disease, as well as more organ damage.[37] Internationally, North America reports the highest incidence of SLE, while the lowest was reported in Africa and Ukraine, and the lowest prevalence was reported in Australia.[38] Thus, there are differences in the incidence and prevalence by location, gender, ethnicity, and age.

When lupus develops in children, although rare, outcomes are generally worse.[27] Surprisingly, lupus has tripled in the last 50 years, and no one understands why.[39] Since lupus is higher in African Americans in the U.S. but lower in people who live in Africa, differences in environment and lifestyles may be a large factor. In 2018, there were 14,263 people newly diagnosed with lupus in the U.S.[32]

Life with Lupus

Ninety percent of SLE patients live with pain, as they suffer from arthralgia (joint pain) and synovitis (painful inflammation of the joint lining).[40] The majority of patients report fatigue as the worst symptom of lupus, with 80-90% reporting that fatigue is the symptom that most negatively impacts their lives.[14] It is a debilitating condition and interrupts daily functioning.[7] Over forty percent of lupus patients reported moderate to severe problems with forgetfulness, while almost one-third reported difficulty concentrating.[41] Beyond the above symptoms negatively affecting daily life, lupus can attack organs and lead to life-threatening illnesses, such as lupus nephritis.

In addition to experiencing the various difficult symptoms and probable multiple organ involvement in SLE, these patients often suffer from additional autoimmune diseases (such as Sjogren's syndrome, Hashimoto's thyroiditis, etc.). The multiple-autoimmune disease pattern is certainly true in my case, as I developed Hashimoto's thyroiditis first (but just thought it was low thyroid), followed by psoriatic arthritis, then lupus and fibromyalgia, and later Sjogren's syndrome and Ulcerative Colitis. Although I have six autoimmune diseases, lupus is the absolute worst!

The specific subject we are discussing in this book is damage to the central nervous system within SLE and, more specifically, damage to the brain, resulting in cognitive dysfunction. "Systemic lupus erythematosus is an autoimmune disease that often causes damage to the skin, kidneys, lungs, heart, and **brain** of affected individuals."[42] To be clear, cognitive dysfunction does not exclusively affect lupus patients but has also been found to plague patients with other autoimmune diseases: Rheumatoid Arthritis (RA), Multiple Sclerosis (MS), etc. Literature regarding cognitive issues in these other autoimmune diseases will also be briefly reviewed toward the end of this book.

Lupus is very complex and not completely understood. The Lupus Foundation of America (LFA) (lupus.org) is a good resource for more general information (see https://www.lupus.org/resources/what-is-lupus), and you can also view basic information here (https://www.cdc.gov/lupus/basics/index.html).

In addition, *The Lupus Encyclopedia* by Donald Thomas (2014) is an excellent, comprehensive resource for SLE patients. If you're looking for an intense and detailed textbook, I recommend *Dubois' Lupus Erythematosus* (Wallace & Hahn, 2007).

In this book, we are focusing on how lupus can damage the brain and cause problems with thinking and memory. Lupus — short for systemic lupus erythematosus (SLE) — is a disease where the body's immune system attacks its own healthy tissues. It often harms areas like the skin, kidneys, lungs, heart, and brain. Problems with thinking and memory aren't only seen in people with lupus. They also happen in people with other autoimmune diseases, like rheumatoid arthritis and multiple sclerosis. Toward the end of the book, we'll also briefly talk about cognitive problems in those diseases.

Lupus is a very complicated illness, and doctors and scientists are still working to fully understand it. In the following pages, I will share everything that I have found through research and personal experience, so read on to enhance your understanding.

Chapter 2

What Are the Treatment Options for Lupus?

Despite decades of medical advancements, the treatment landscape for lupus remains limited, inconsistent, and, in many cases, inadequate. Sadly, there is no cure for lupus and no available treatment helps everyone. In fact, there are very few treatments, and only three FDA-approved medications have been developed specifically for lupus. Did you hear me say that there have been only three treatments in the last 60 years of medical breakthroughs? What's the reason—or the excuse—for the lack of available treatments? I'm unsure, as I've been wondering this myself for over a decade. I know lupus disease is complicated, and it affects each patient in different ways, but I think it is inexcusable that there are so few treatment options! What's worse, "the lack of a curative therapy leaves most patients with a long-term sickness, which can negatively affect their emotional, psychological and social functioning and quality of life."[1]

Remember, SLE can damage multiple organs in the body. It is "characterized by a rupture of self-tolerance and systemic inflammation mainly resulting from the hyperactivation of peripheral B and T cells, resulting in high levels of pathogenic autoantibodies, tissue deposition of immune complexes, and, ultimately, **multiple and various organ injuries**."[2]

This chapter includes the current approaches to lupus treatment, including long-standing therapies such as antimalarials, corticosteroids, and NSAIDs, as well as more recent biologics and off-label immunosuppressant use. Each category of treatment offers benefits but also presents significant challenges, including toxicity, tolerability issues, and long-term risks. Additionally, while some medications may help control disease activity, very few lead to full remission, and many patients continue to experience residual symptoms that interfere with daily life.

Overview of Lupus Management

Although treatment for lupus is not uniform, hydroxychloroquine (originally given for malaria) is the most common long-term treatment. A recent article summarizes four main types of management for SLE:
    1) Lifestyle changes such as avoiding ultraviolet light
    2) Prevention of comorbidities (osteoporosis, heart disease, infections)
    3) Immunomodulator drugs (like hydroxychloroquine)
    4) Immunosuppressants (like Benlysta)[2]

Again, there is no cure. One of the key challenges in managing lupus lies in addressing its frequent overlap with other health conditions. Comorbidities (additional diagnoses) are common in autoimmune diseases. In fact, I have yet to meet a lupus patient who is dealing with "only" lupus, although that would certainly be enough of a burden. A recent article reporting the real-world burden of lupus identified many comorbidities, including asthma, hypertension, arthritis, and heart disease, that were greater in the lupus population as compared to those without lupus.[3] The complications surrounding having several diagnoses are important to remember when searching for appropriate treatment solutions.

Although there are very few medications approved specifically for lupus, advances in treatment over several decades have improved patient survival rates. Despite the improvement in survival rates, however, lupus management still does not provide complete remission for patients, and even those considered to be in remission by their physicians still have a host of residual symptoms that remain as part of their daily lives.[4]

Antimalarials

Hydroxychloroquine, aspirin, and corticosteroids were approved in the 1950s but were not designed for lupus management, and the approval was not based on clinical trials but on clinical experience.[5] Hydroxychloroquine is commonly prescribed long-term for lupus patients to attempt to prevent flares. In addition, antimalarial drugs like hydroxychloroquine have been found to reduce overall organ damage and decrease mortality rates, so these are prescribed in almost all SLE patients.[6] Further, antimalarials have the added benefit of being negatively associated with neuropsychiatric events.[7]

On the downside, hydroxychloroquine (brand name Plaquenil) can cause retinal toxicity, particularly in patients who have taken the drug for over ten years. Annual screening tests by ophthalmologists are recommended to monitor changes in retinal health. Another possible side effect of antimalarials (and one I was entirely unaware of) is an increase in myopathic toxicity (causing muscle pain and weakness) in skeletal and cardiac muscles with long-term use.[8] Due to its anti-inflammatory effects, some researchers think hydroxychloroquine may decrease the risk of cognitive impairment.[9] However, this has not yet been proven.

Moreover, some lupus patients can't tolerate hydroxychloroquine, so the anti-malarial drug Aralen (chloroquine) may be used. Chloroquine has a reputation for more significant side effects, as compared to hydroxychloroquine.

Steroid Use

Steroids such as prednisone and methylprednisolone have been used in as high as 90% of lupus patients despite adverse side effects related to their use.[10] Even small doses, when taken over a long period of time, show an increased morbidity.[11] Furthermore, it also causes long-term consequences such as organ damage and osteoporosis.[6] Despite the side effects, however, they are effective at managing the pain and inflammation associated with lupus, and they must be utilized in moderate to severe cases, at least on a short-term basis.

It is advised to reduce the dose as soon as possible and to wean completely off steroids when possible. However, despite this goal, a recent study analyzed medication treatment trends of SLE patients from 2007 to 2023 and found that 66.7% of lupus patients in the U.S. continue to rely on glucocorticoids (steroids).[12]

NSAIDS

Non-Steroidal Anti-Inflammatory Drugs (NSAIDS) are commonly used in lupus patients to reduce inflammation. Some are over-the-counter, like Ibuprofen and Naproxen, while others are prescription-based, like Celexocib and Relafen. Aspirin is the most common NSAID and is often used to prevent blood clots.

However, despite their accessibility and effectiveness, NSAIDs are not without drawbacks. They can cause stomach pain, ulcers, and kidney damage with long-term use. I personally experienced severe stomach pain and gastritis (inflamed stomach).

New Medications/Biologics

For decades, patients had limited options specifically designed to target lupus, often relying on broad-spectrum immunosuppressants and corticosteroids. However, in the last decade, significant progress has been made with the development and approval of medications tailored to the underlying mechanisms of the disease. This marks a new era in lupus management, providing more precise and effective treatment options.

- **Benlysta (Belimumab):** The next medication to be approved for SLE appeared almost sixty years later. The new drug–belimumab (Benlysta)–was designed for lupus and released in 2011. As of 2011, Benlysta was the only medication specifically developed to target generalized SLE. Benlysta (name brand) or belimumab is a monoclonal antibody medication that acts during B cell division and prevents the B cell-activating factor (BlyS) from binding to cells. Benlysta stops cell division and growth.[13] Abnormalities of B cells and plasma cells are thought to be one main factor in lupus development.[14] Benlysta is well-established now as a frequent option for the treatment of moderate to severe disease and is believed to be safe for longer-term therapy (at least up to seven years).[15] It has had limited success in many patients; however, so this highlights the need for different types of treatments.[16]

- **Saphnelo (Anifrolumab-fnia):** In 2021, the FDA approved anifrolumab-fnia (brand Saphnelo) for moderate to severe SLE. It was the first medication approved in a decade for general lupus. Saphnelo is a Type I interferon receptor antibody medication. Since Type I interferon (a cytokine that plays essential roles in inflammation & immunoregulation) is thought to play a role in the pathophysiology of lupus, this new treatment is encouraging. The upregulation of interferon-stimulated genes has been found in 60-80% of lupus patients, which may represent a subgroup of lupus patients with more severe disease and activity in multiple domains.[17] Saphnelo "dampens the excessive type I interferon signature found in up to 80% of adults with lupus."[18] Furthermore, research is being completed to determine if Saphnelo will also be effective for lupus nephritis and/or CNS lupus. Saphnelo and Benlysta can both cause nausea, diarrhea, fever, sleep interruption, pain in limbs, migraines, and increased infection.
- **Lupkynis (Voclosporin):** It was approved in January 2021 for treating lupus nephritis. Lupus nephritis is a common complication of severe lupus, in which serious kidney damage and kidney failure can occur. Belimumab was also recently approved for the treatment of lupus nephritis.

Altogether, three medications—Benlysta, Saphnelo, and Lupkynis—have been approved for adult SLE in the past decade, marking significant progress in treatment. Interestingly, in a recent review of 2476 lupus patients and their medication usage, adherence to their prescribed meds was highest for biologics, followed by antimalarials and immunosuppressants, but indicated that corticosteroid use was also high.[19]

Off-Label Immunosuppressant Use

Other immunosuppressant medications are used off-label for lupus. These include medications like cyclophosphamide (Cytoxan), azathioprine (Imuran), mycophenolate (Myfortic), and methotrexate (Reumatrex).[13] Mycophenolate mofetil (MMF) is another immunosuppressive drug that reduces tissue inflammation.[20]

In addition to this, a recent study compared two immunosuppressive drugs and found that sirolimus (Rapamune) was more effective than tacrolimus (Prograf) in 52 patients with lupus.[21] These immunosuppressant medications work by interfering with DNA in the nucleus of immune cells. Immunosuppressants are chosen based on patient characteristics: disease severity, organ manifestations of disease, patient age, child-bearing, etc.[22]

Personal Medication Experience

My personal journey with medications for lupus has included several of the above drugs. I had been on hydroxychloroquine for 16 years until recently. I didn't notice any side effects, except maybe nausea. As mentioned, eyesight must be carefully monitored each year for damage. Early on in my disease, I was put on methotrexate when I had proteinuria (protein in the urine). It definitely helped with that, but it was a difficult medication to tolerate, causing extreme nausea, hair loss, and weight loss. I have had oral corticosteroids many times for various symptoms (pleurisy, arthritis) and numerous steroid injections into specific joints. I tried Imuran for a while but didn't see results—only side effects. As mentioned previously, Benlysta helped some with symptoms and possibly stopped the rapid progression of the disease, but it did not ultimately resolve cognitive issues, fatigue, pain, or fever.

For me, monthly Benlysta infusions were each followed by a week of extreme fatigue (much more than normal), with three days in bed after an infusion each month (suffering from migraine, body aches, dizziness, nausea, exhaustion, insomnia). In addition to all those symptoms, I had significantly increased cognitive dysfunction for 2-4 days after each infusion, which caused me to worry about long-term brain damage with continued use.

Besides those, I also took Humira injections for years for psoriatic arthritis, which helped with the pain but caused me to have frequent sinus infections. I also took Celebrex for many years, as well as prescription Ibuprofen. Those meds damaged my stomach and microbiome. I am thankful for the available medications, but we do not have enough options. We need better treatments with fewer side effects. In fact, side effects from medications often cause progressive damage in the body, which contributes to increased mortality.

Further, we need treatments that do not suppress the immune system, making us more vulnerable to infections. Hopefully, treatments such as peptides that are non-toxic with no side effects can eventually replace the current toxic, immune-suppressing drugs (see Datta, 2021,[23] for more information on this emerging treatment). In many cases, certain lifestyle changes can replace the need for serious medications, which is what happened to me.

Treat to Target

The latest approach for clinicians is called "treat-to-target," in which the goal is to minimize disease activity.[23] "Treat-to-target" must be individualized and may vary from patient to patient. The main goal is "absence of activity in all organ systems (reflected in a SLEDAI close to zero) with the lowest possible dose of CS (steroids) and standard, non-toxic maintenance of immunosuppressant drugs."[23] Unfortunately, the symptom of cognitive dysfunction is often ignored in standard treatment. We must remember that "dismissing cognitive dysfunction as merely a confounding symptom in SLE is shortsighted and does a disservice to patients."[24]

Furthermore, it has become apparent that the definition of remission in SLE is not uniform across clinicians or countries. To address such variability, the Definitions of Remission in SLE (DORIS) Initiative was formed to attempt to provide agreement on a definition of remission.[25] They have acknowledged that remission off treatment (medication) is very rarely achieved. Thus, they set guidelines for remission with certain scores on two clinical scales SLEDAI & PGA (Physician Global Assessment) but allow the patient to remain on medications while still achieving their definition of "remission."[25]

An additional scale, the British Isles Lupus Assessment Group (BILAG), is a different organ-based measure utilized in some countries.[26] As a patient, I find allowing for "remission" while on medications problematic, as I feel true remission would be a healthy body requiring no medications. Since medications are often necessary, however, I am thankful there are more targeted lupus medications currently in clinical trials, so hopefully, there will be safer options for the treatment of this disease and, particularly, the resulting cognitive impairment in the near future.

In summary, the current treatment landscape for lupus is marked by significant progress but also considerable challenges. While medications like hydroxychloroquine, corticosteroids, NSAIDs, and newer biologics such as Benlysta, Saphnelo, and Lupkynis offer hope, they come with side effects and limitations. The overall treat-to-target approach aims to personalize treatments to achieve minimal disease activity and improve quality of life. However, the complexity of lupus and its interaction with other autoimmune disorders make management a constant balancing act.

Continued research and development of more targeted therapies promise a brighter future for those living with this debilitating disease. It is crucial to remember that lupus management must evolve to address both physical and cognitive symptoms, ensuring holistic care for patients. It is also important to remember that diet and lifestyle can make a huge difference in many cases (more on this to follow).

Chapter 3

What Causes Lupus?

Understanding what causes lupus is one of the most pressing questions in autoimmune research. Although there is no single answer yet, experts agree that lupus is a multifactorial disease shaped by a complex interaction between genetic predisposition, immune system abnormalities, and environmental triggers.

In this chapter, we'll explore the current scientific insights into the causes of lupus, from inherited genetic variations to immune system dysfunction, hormonal influences, and the role of environmental factors. While researchers continue to piece together the full picture, what we do know is steadily guiding us toward better prevention, earlier diagnosis, and more targeted treatment options.

Unraveling the Mystery Behind Lupus Causes

The etiology (cause) of lupus is still under investigation. However, experts believe it to be multifactorial and complex, such that there is a genetic predisposition for lupus, and then an environmental trigger (or triggers) sets it off. Sometimes, this complicated relationship is discussed as genetic factors and epigenetic alterations.[1] This view is supported by a renowned group of scientists who state that "human SLE is believed to be triggered by environmental factors in genetically susceptible individuals."[2]

The Genetic Complexity of Lupus

We know now that lupus has a genetic component, although it is a non-Mendelian disorder (not linked to single genes).[3] Moreover, eight genes related to SLE were examined.[4] The genetic component of the disease is extremely complex, as over 80 genomic loci (specific positions on a chromosome) have been associated with SLE, which "lead to the formation of key proteins, each contributing a small increase to the risk" of lupus.[5]

By 2020, genome-wide association studies had identified over 90 SNPs (single nucleotide polymorphisms or variations in position in a DNA sequence) associated with SLE. The "main genetic risk loci for lupus susceptibility in GWAS (genome-wide association studies) are located in MHC class II and IRF5 regions, which respectively determine autoantigen presentation and associated activating cytokines production required to recruit autoreactive T helper cells."[6]

Emerging Discoveries in Lupus-Related Genes

Recent research continues to uncover additional genetic contributors. One study found 7 proteins to be associated with SLE and possible novel genes associated with the disease.[1] Interestingly, researchers have found that lupus may share genetic differences with the disease of macular degeneration.[7]

An additional study provided "clear evidence" that a TLR7 genetic mutation on the X chromosome in the DNA of a seven-year-old with severe lupus was the cause of her lupus.[8] Such research reveals that there are indeed genetic components, but it is generally accepted that "genetic predisposition, environmental triggers, and the hormonal milieu interplay in disease development and activity."[9]

Rethinking Lupus as a Single Disease

Given the wide range of genetic abnormalities observed in lupus patients, some investigators have even begun to question whether SLE is a single disease, citing various predisposing genetic defects.[10] Since "no single gene (or combination of genes) nor the spectrum of exposures (to an environmental factor) of an individual" has been determined to be the cause of lupus, some researchers propose the examination of gene-environment interaction factors in case studies, to determine possible subtypes of lupus pathogenesis.[11] Regardless of the details, most scientists now accept that environmental and genetic factors somehow interact to trigger the disease.

Genetic Risk Among Family Members

Since genetics has a role in lupus, relatives of those with lupus are more likely to also develop the disease. In fact, a nationwide cohort study in Denmark revealed that risk was highest among first-degree relatives but still present among second and third-degree relatives.[12]

Interestingly, the increased risk wasn't only for lupus, as relatives of lupus patients were also more likely to develop other autoimmune diseases, such as rheumatoid arthritis and inflammatory bowel diseases. A recent study found that systemic inflammation and immune dysregulation were the two main factors influencing whether a relative developed lupus.[13] From this, I would conclude that actively working to keep inflammation down and the immune system balanced (anti-inflammatory diets, stress management, hormone regulation, adequate sleep, exercise, etc.) would be a primary goal for prevention of this disease.

Recognizing Early Warning Signs

Cognitive dysfunction, as well as headaches, anxiety, and fatigue, may even be an early sign of the developing disease.[13] In fact, researchers have found that using the lupus portion of the Connective Tissue Disease Screening Questionnaire (SLE-CSQ) can help identify early signs of lupus.[14] More information like this is of utmost importance in working toward the prevention of autoimmune diseases and is a personal quest for those of us with children and grandchildren.

Multifactorial Disease

As mentioned, no one definitively knows what causes lupus, but it is generally recognized as a multifactorial disease with genetic, immune, and environmental factors.[15] Regarding immunology, there are abnormalities in cellular and humoral immunity.[16] These abnormalities are caused by two specific types of immune cells (lymphocytes) that have been linked to lupus: B cells and T cells. B cells have been found to play a role in the pathogenesis of lupus, as B cell activation produces autoantibodies.[17] Other cells are supposed to regulate T cells and keep them from harming the body, but these regulatory cells fail in lupus, which "allows T cells to go rogue and attack the body, causing the symptoms of lupus."[18]

It is thought that the activated T cells secrete proinflammatory cytokines, which play a role in the pathophysiology of lupus through systemic inflammation.[19] Cytokines carry signals locally between cells and are secreted by immune system cells, playing a vital role in both innate and adaptive immune responses.[20]

Moreover, a strong association has been recently found between lupus and Treg cell defects, and Treg cells are vital to providing tolerance for autoimmunity and preventing abnormal activation of the immune system.[21,22] Forkhead box P3 (FOXP3)-expressing Treg cells have been found to play a significant role in the regulation of immune response and the maintenance of immune homeostasis, such that researchers are exploring how to manipulate these cells in humans as a treatment for autoimmune disease, as well as cancer.[23] A recent mouse study concluded that increased FOXP3 E2 deletion isoform expression leads to the interruption of immune balance and is associated with autoimmune diseases.[24] IL-2 affects Treg cells, so low-dose IL2 is being studied for lupus patients as a treatment.[25]

In addition to B cells and T cells, ninety-five percent of patients with SLE have antinuclear antibodies (ANA) in the blood, and that is one standard test that is used for diagnosis and then continuously monitored in all lupus patients.[17] Autoantibodies attack cell nuclei, which cause numerous problems in lupus.[26] Autoantibodies can even initiate a cytokine/chemokine (chemokines are cytokines that play key roles in the accumulation of inflammatory cells) storm.[20] A group of researchers discuss the loss of both normalized innate and adaptive immunity, with reference to antinuclear antibodies, B cells, T cells, cytokines, and apoptosis.[27]

Reactive Oxygen Species

Another possible factor in lupus development is the body's inability to remove reactive oxygen species. Reactive oxygen species (ROS) are produced in the body as part of normal cellular processes, but the body should be able to clear them out. Prolonged exposure to ROS can cause oxidative damage to DNA and has been tied to fatigue in lupus.[17]

Furthermore, some researchers believe that oxidative stress and mitochondrial dysfunction play a big role in the pathogenesis of lupus. A study found decreased amounts of mtDNA (mammalian mitochondrial DNA) in lupus patients and increased mtDNA damage with greater disease duration and damage.[28] The ROS mentioned above can damage the DNA and lead to decreased function of the mitochondria.

More recently, a study showed that programmed mitochondrial removal is defective in SLE patients, causing an accumulation of red blood cells carrying mitochondria.[29] This process, in turn, leads to high levels of blood IFN-stimulated gene signatures. They concluded that "emerging evidence supports that mitochondrial dysfunction contributes to systemic lupus erythematosus pathogenesis."[29]

Related to this, a recent study discovered healthy adults benefited from taking a nicotinamide mononucleotide (NMN) supplement to increase levels of nicotinamide adenine dinucleotide (NAD+), which directly influences mitochondrial function, and the author stated this supplement may help with immune function, as well as cognition.[30]

Environmental Factors

Experts believe that "SLE may be triggered in a genetically-susceptible individual by exposure to certain environmental risk factors."[31] Certain factors in the environment cause immunologic responses, which can lead to inflammation and autoantibody production, causing tissue damage. Ultraviolet light and specific medications are known environmental triggers.[32]

However, others (such as certain infections or suboptimal nutrient levels) are still under investigation. It is thought that "multiple mechanisms lead to immunological abnormalities."[28] Some researchers have found an association between lupus onset and viral infections, such as human endogenous retroviruses or Epstein-Barr virus.[33] These viruses (or possibly bacteria or fungi) then trigger an overactive immune response through dendritic cells, IFN, T-cells, and B-cells.[33]

Type I Interferon and Microglia

An article summarized a few theories of SLE pathogenesis: type I interferon may play a role by disrupting peripheral immune tolerance; IFN-alpha with complement cascade may cause synaptic pruning through microglia, or chronic inflammation may activate microglia to stimulate synaptic loss.[34] Type I interferons normally provide acute defense in the body in response to a virus. When the production of type I interferons continues for too long, however, it can change the immune system. "Studies of lupus patients have supported the conclusion that type I IFNs comprise a family of pathogenetic mediators that contribute to autoimmunity, inflammation and ultimately tissue damage in patients with SLE."[3] In fact, SLE has been called the "archetype" of "acquired autoimmune interferonopathies," as the adaptive immune system is activated and causes inflammatory responses.[35]

Further, the IFN signature (evidence of activation of the IFN system through increased expression of sets of IFN-regulated genes) was detected in lupus over 20 years ago.[35] In one recent study, the "frequency of IFNalpha-producing monocytes positively correlated with SLE disease activity."[36] It is thought that "a primary disturbance of the innate immune system can spill over into autoimmunity," but "autoinflammation can be both IFN and non-IFN related."[3] Interestingly, type one interferon has been found to be associated with decreased memory and cognition in Alzheimer's disease.[37] So, there may be a connection there with cognitive disturbance in lupus.

Microglia, the brain's resident immune cells, are another critical piece of the puzzle in SLE. Studies suggested that "the JAK/STAT3 pathway is a universal signaling cascade involved in virtually all pathological situations in the brain" and relates this pathway to microglial cells, cytokines, reactive astrocytes, and neuroinflammation.[38]

Turning to the theory of immune tolerance, Thurman & Serokova state that SLE "is associated with a loss of immunologic tolerance to multiple nuclear antigens and the production of autoantibodies specific for these self-antigens."[39] Thus, the treatment almost always requires immunomodulatory therapies. Although specific theories differ, it is clear that the immune system is different in patients with lupus. In a recent study analyzing the immune profiles of lupus patients vs. healthy controls, researchers found a distinct pattern of lower dendritic and natural killer cells and higher CD8+T cells in the SLE patients.[40]

In addition, recent research also found that monocytes act differently in people with lupus than they do in healthy controls.[29] Monocytes are a form of white blood cells that fight off bacteria, viruses, etc., when needed. They normally express interferon-sensitive proteins or interleukin-1 beta in healthy individuals, but they co-expressed both in lupus patients.

To sum up, researchers hope to expand on this to create new treatments that target interleukin-1 beta since current treatments targeting interferon do not work for everyone with lupus. Moreover, immune dysregulation is finally becoming clearer in lupus, although we have a long way to go in understanding this complex disease.

The Microbiome Factor

A very new, interesting theory of lupus pathogenesis stems from studying the microbiome, which plays a huge role in human health.

The human microbiome consists of the "10-100 trillion microbial cells that include bacteria, yeast, archaea, protozoa, and their infecting viruses."[41] Most people are somewhat familiar with the microbiome due to the popular use of probiotic supplements. The microbiome plays a role in balancing inflammatory and regulatory responses.[22]

Impact of the Microbiome
The microbiome's composition can affect the immune system, particularly T helper and T regulator cells; less diverse microbiota has been tied to SLE and can result from a high-fat diet and/or obesity.[42] In fact, "the intestinal microbiome has been proposed as a link between dietary patterns and development of autoimmunity."[43]

Some researchers have discovered that the function and composition of the microbiome of lupus patients are very similar to that of aging individuals.[44] This microbiome finding led scientists to conclude that lupus may be similar to the pathophysiological process of normal aging or immunosenescence (gradual immune system deterioration). Premature biological aging, or senescence, may occur in lupus and then be related to dysregulation of both the innate and adaptive immune systems.

Dysbiosis and Its Role in Chronic Inflammation
Immune system changes would cause chronic inflammation and dysbiosis (imbalance of good and bad bacteria in the microbiome, as well as reduced diversity of species) and intestinal permeability (leaky gut). The dysbiosis and leaky gut allow autoantibodies to be created, which attack healthy tissue in both lupus patients and normally aging individuals. Chronic low-grade inflammation is common with dysbiosis and intestinal permeability.[45] Moreover, such permeability can be detected through the biomarker Zonulin.[46]

Further, researchers have discovered a decreased amount of Firmicutes phylum bacteria and too much Bacteroidetes, which then leads to decreased production of short-chain fatty acids, like Butyrate. Short-chain fatty acids are vital to maintaining the intestinal barrier (thereby preventing leaky gut). Greater consumption of these in a diet can help maintain intestinal integrity, which prevents bacteria from entering the bloodstream.[43]

In addition, short-chain fatty acids help reduce proinflammatory cytokines and increase anti-inflammatory cytokines. A recent study found that a class of anti-inflammatory lipids known as SGDGs decreases in aging, allowing lipopolysaccharides (LPS – dangerous endotoxins) to flourish and trigger inflammatory cytokines in the body.[47] As we know, a proinflammatory environment is definitely present in lupus. "Manifestations of lupus and senescence stem from similar features of dysbiosis in both aging individuals and lupus patients, which create the proinflammatory environment leading to 'inflammaging' and immunosenescence in these populations and may be the cause for similar clinical features seen in both populations."[47]

Gut Microbiota Alterations and Lupus Flares

The authors then pose the possibility that the microbiome might be a target for treatment for both lupus patients and diseases of normal aging (infections, cancers, cardiovascular disease, etc.). There is a "potential role for microbial triggers in driving the immune system activation that leads to disease flares."[3] Another study found significant alterations in the gut microbiota of lupus patients, as well as decreased diversity of the microbiome, and was even able to find a pattern of microbiota distribution.[48] Certain strains were elevated in lupus patients, while others decreased compared to healthy controls. It has also been postulated that a distinct profile of disturbed gut microbiota in lupus patients may lead to the production of inflammatory cytokines.

Microbiome as a Therapeutic Target

Another study found a reduced Firmicutes/Bacteroidetes ratio in SLE patients, with researchers concluding that administering specific probiotics to lupus patients could potentially provide therapeutic benefits by restoring the Treg/Th17/Th1 imbalance.[22] This means scientists noticed that people with lupus (SLE) had a different balance of gut bacteria compared to healthy people. These bacteria groups—Firmicutes and Bacteroidetes—are normally found in the gut, and their balance is linked to a healthy immune system. When that balance is off, as in lupus patients, it can affect how the immune system works. The researchers think giving the right probiotics (helpful bacteria) could help fix this and possibly reduce lupus symptoms by calming the overactive immune responses. A recent study has found an association between the pathobiont Enterococcus gallinarum and anti-human RNA autoantibody responses in SLE patients.[49] In simpler terms, scientists found a potentially harmful bacteria (*Enterococcus gallinarum*) in some lupus patients that might be encouraging the immune system to attack the body's own RNA (an essential messenger molecule). This connection helps us understand one of the possible ways lupus symptoms could start or get worse, but more research is needed to confirm the link.

Interestingly, probiotics were recently found to improve cognition in healthy adults.[50] Lupus nephritis flares, specifically, were associated with the expansion of Ruminococcus gnavus in the microbiomes of half of the patients studied.[51] Simply put, when lupus causes inflammation in the kidneys (called lupus nephritis), it often comes with a rise in a certain gut bacteria called *Ruminococcus gnavus*. This suggests that the worsening of kidney symptoms might be linked to what's happening in the gut, highlighting again how closely gut health and immune problems are connected. Furthermore, the oral microbiota has also been found to be disturbed (oral dysbiosis) in lupus patients.[52] Research is needed to determine if oral dysbiosis leads directly to gut dysbiosis, although that connection was certainly obvious in my personal testing.

To summarize, more studies on the microbiome are needed since "while studies on autoimmune rheumatic disease have almost invariably shown abnormal microbiome structure dysbiosis, substantial variability in microbial composition between studies makes generalization difficult" and "an etiopathogenic role of specific pathobionts cannot be inferred by association alone."[41] Basically, most studies agree that people with autoimmune diseases like lupus have unusual gut bacteria. However, the exact types of bacteria that are affected vary a lot from study to study. So, it's still hard to say for sure which bacteria are causing problems and which ones are just innocently present in the gut. More research is needed to make strong conclusions.

In an attempt to pull the findings of microbiome studies together, a group of researchers did a meta-analysis of 11 case-controlled studies encompassing 373 lupus patients compared to healthy controls.[53] They found definite associations between the microbiome and immune systems, and that ruminococcaceae, enterococcaceae, and enterobacteriaceae species were all imbalanced in the gut of SLE patients.[53] To get a clearer picture, scientists looked at the results of 11 different studies and compared gut bacteria in lupus patients with healthy people. They found patterns showing that some bacteria families—like ruminococcaceae and others—were consistently out of balance in people with lupus. This supports the idea that gut bacteria are closely tied to how the immune system behaves in lupus.[53]

In addition, they determined that the common lupus medications corticosteroids and hydroxychloroquine altered the microbes in the gut. Other researchers have found that skin microbiota differences may play a role, as well as in the case of psoriatic arthritis.[54] In fact, one study states that oral, gut, and skin microbiota may all play a part in contributing to lupus, as they "hypothesized that autoimmune-predisposed individuals with the right combination of genetic susceptibility and environmental exposures who are chronically colonized by Ro60 commensals may develop antibodies against a bacterial Ro60 ortholog that leads to autoimmunity via cross-reactivity and epitope spreading."[55] They suggest that bacteria already in and on the body (normal flora) may initiate and drive lupus pathogenesis but trigger chronic autoantibody production.

Therefore, gut dysbiosis is now thought to play a crucial role in most, if not all, autoimmune diseases.[56] Interestingly, dysbiosis of the gut is tied to ultra-processed food (UPF) consumption, and higher UPF daily intake was found to increase the risk of lupus by over 50%.[57] So, particular diets are likely one contributing factor to this disease.

Accelerated Immune Aging

Returning to immunosenescence, accelerated immune aging has indeed been found in SLE patients. An immune-risk ratio was calculated in a study and found to correlate with cognitive impairments, as measured by the MMSE and MoCA, and self-reported as "lupus brain fog."[58] It suggests further research into T cell senescence profiles for lupus patients as a possible biomarker to measure and track cognitive changes in the future.

The senescence of cells is a primary hallmark of aging and age-related disease.[59] A recent study of immune dysregulation in the CSF (cerebrospinal fluid) of individuals revealed that those with cognitive impairment had patterns of much older individuals.[60] I think investigating premature cellular senescence in lupus is an interesting area of research and one that makes sense in light of lupus symptoms: cognitive impairment, fatigue, and joint pain, which all sound like complaints of an older person. While currently at age 53, I may be bordering on "old," but I remember suffering significantly from lupus in my early 30s and thinking I felt like a much older person, and even noticing that I had less energy than my own parents! Still, while immunosenescence may be a causal factor, it is not likely the only factor.

Since the etiology of lupus (SLE) is multifactorial, the etiology of cognitive impairment in lupus is probably also multifactorial, and the combination of factors causing the disease likely differs for each person with lupus. Partial causality may be attributed to the microbiome and immunosenescence, as well as to genetics and suboptimal nutrient levels in one patient. Another patient may have different genetic SNPs (single-nucleotide polymorphisms-variations in the genome), chronic hidden infection, and mold toxicity. In the best-case scenario, then, a team of providers (rheumatologist, functional medicine physician, health coach, etc.) would work together, utilizing the guidance of specific test results, to determine the specific causal factors for that one patient, and develop an individualized plan for how best to support that patient's health. The goal is to help bring the body back into balance, which may include a combination of medications, supplements, nutrition, and other lifestyle adjustments and therapies.

In conclusion, the causes of lupus are profoundly complex and multifactorial, involving a dynamic interplay between genetic, immune, hormonal, environmental, and even microbiome-related factors. Research has revealed significant genetic risk loci, immune system abnormalities, and environmental triggers that collectively set the stage for disease development. Emerging discoveries around mitochondrial dysfunction, oxidative stress, and gut dysbiosis have further expanded our understanding, highlighting the intricate biological processes behind lupus. Although lupus is not caused by a single gene or exposure, the growing body of evidence suggests that the interaction between these many elements leads to the characteristic immune system dysregulation seen in the disease.

Advances in genetic studies, immune profiling, and microbiome research continue to shed light on the pathways that contribute to lupus onset and progression. These insights are paving the way for more precise diagnostic tools, targeted treatments, and possibly even preventive strategies in the future. Despite the complexity, the scientific community is steadily moving closer to unraveling the full picture of lupus pathogenesis.

Ongoing research will be critical in further clarifying these relationships and, ultimately, improving the lives of those affected by this challenging and often misunderstood autoimmune condition.

Chapter 4

What is Cognitive Impairment?

Living with lupus often means navigating a world of invisible symptoms, and for many of us, "brain fog" is one of the most frustrating. It's a term that resonates deeply with patients and clinicians alike in everyday conversation, yet it rarely appears in research literature in the same form. Instead, it's replaced with more clinical terms like cognitive impairment, cognitive dysfunction, or mild cognitive impairment (MCI).

As both a lupus patient and a speech-language pathologist, I've experienced this disconnect firsthand—feeling the very real effects of cognitive challenges in daily life while also recognizing the vast deficits I used to treat in others. This chapter explores the gap between the subjective experience of "brain fog" and how it is defined, measured, and addressed in research and clinical settings. My goal is to bridge that gap by using both personal insight and professional knowledge to shed light on what these cognitive struggles really mean, how they affect us, and why they deserve more serious attention.

Understanding Brain Fog in Lupus

There seems to be a disconnect between patient/clinic life and research, as "brain fog" is acknowledged as a common issue in lupus. Still, in the research literature, it is defined as cognitive impairment, cognitive dysfunction, or MCI (Mild Cognitive Impairment). As mentioned in the introduction, I am a lupus patient who has suffered from inconsistent cognitive deficits. In addition to that, I am a speech-language pathologist (SLP) who has evaluated and treated cognitive impairment (resulting from various diseases/disorders) in the past.

In exploring lupus-related content, I found that internet articles about lupus, patient blogs, and even rheumatologists talk about "brain fog" as a common, bothersome symptom of lupus. From the readings, it seemed like most people accepted it as an annoyance to be dealt with, but I despised having to deal with this "fog." It seriously affected my life, my work, and my self-confidence. I understand why they call it "fog," as it sometimes felt like my head was full of thick fog, and I was slowly trudging or pushing through my brain to reach a word or conclusion. Since that time, I've worked with other autoimmune patients with similar descriptions of brain fog, and they've revealed that they are frightened and worried about their futures, since the cognitive changes affect every aspect of life.

One day, I was thinking through my specific symptoms of difficulty with word-finding, slower processing rate, decreased reading comprehension, and poorer working memory when I realized that these were serious, measurable cognitive deficits! I then took a couple of online cognitive screeners to confirm my fears. Although on a lesser scale, I was dealing with some of the very same issues that I had addressed in treatment with my former patients. I wondered why my doctors hadn't seriously acknowledged my cognitive deficit when I brought it up and hadn't offered me a referral for evaluation and treatment when I complained about these difficult symptoms that were negatively affecting my daily life.

What is Cognition?
What is cognition, and what does it entail? In the simplest terms, cognition is your ability to think. "Cognition is the sum of intellectual functions that result in thought. It includes reception of external stimuli, information processing, learning, storage and expression."[1] Cognition is directly tied to the brain, and the human cortex has 30 billion neurons and can make one million billion

synaptic connections, so the "human brain can be described as the most complex known object in the universe."[2] As a result, cognition and cognitive impairment are both extremely complex and not completely understood.

What is Cognitive Impairment Like?

I've alluded to what cognitive impairment was for me, but it can vary from person to person. According to Mackay, "Lupus brain fog is an extremely common patient complaint that refers to periods of forgetfulness and confusion that are related to impaired cognition."[3] In an interview, a lupus patient with brain fog said that "it can really make your whole world fall apart."[4] The Lupus Foundation of America explains brain fog as having the following deficits: "the difficulty that you may have in completing once-familiar tasks such as remembering names and dates, keeping appointments, balancing your checkbook or processing your thoughts."[5]

Essentially, it can be thought of as a cognitive impairment that is milder than dementia. In fact, a study points out that the concept of mild cognitive impairment is twenty years old and was initially created to help research and clinical practice find people at risk for dementia.[6] MCI can be defined as 'the symptomatic *predementia* stage on the continuum of cognitive decline, characterized by objective impairment in cognition that is not severe enough to require help with usual activities of daily living."[7] Furthermore, research specifically states that MCI "encompasses attention, concentration, memory, comprehension, reasoning, and problem solving."[8]

However, they go on to state that MCI can be stable or fluctuating and may even be reversible. Even at the mild level, cognitive deficits can have an impact on daily function.[9] More specifically related to lupus, the ACR (American College of Rheumatology) defined cognitive impairment (1999) in NP (neuropsychiatric)-SLE patients as "significant deficient functioning in at least one of the following cognitive domains:

simple or complex attention, learning and memory, visuospatial processing, psychomotor speed, verbal fluency, reasoning ability, problem solving, and executive processes of planning, organization, and sequencing."[10]

The diagnosis of NP+SLE is difficult, as "there are no unequivocal clinical parameters or definitive laboratory tests for the diagnosis of NP+SLE."[11] NPSLE used to be known as lupus cerebritis.[12] NP+SLE can "involve both the central and peripheral nervous systems, and is the second leading cause of mortality and morbidity in systemic lupus erythematosus patients."[13] Cognitive impairment is one type of manifestation of NP+SLE. On a broad scale, "cognitive function is a clinical surrogate of overall brain health, with applications in both diagnosis and determination of clinical outcomes."[14]

Does Everyone Have It?

Does everyone with lupus, then, have cognitive impairment? That is still under debate. A meta-analysis found that people with SLE (regardless of overt neuropsychiatric involvement) have statistically significant deficits in visual attention, cognitive fluency, immediate visual memory, and visual reasoning.[15] Now, if you haven't read many journal articles or taken a statistics course, you may be wondering what a meta-analysis is. It is a type of statistical analysis that looks at the results of many scientific studies and combines those results into one paper, essentially summarizing what's out there in the literature.

That same study found that patients diagnosed with NP+SLE had greater cognitive deficits than those patients with plain SLE (or NP-SLE).[15] Another meta-analysis found that NP+SLE patients (those with neuropsychiatric SLE diagnoses) consistently "showed poorer cognitive performance and greater cognitive impairment than NP-SLE patients, at least in the areas of visuomotor coordination, attention, executive function, visual coordination, visual learning/memory, and verbal fluency."[16] However, the two groups did not differ significantly in the areas of verbal learning/memory, cognitive flexibility, and inhibition.

This disconnect isn't a new discovery. As early as 1992, a group of renowned researchers noted "marked cognitive impairment" in a significant percentage of SLE patients, even without overt evidence of central nervous system (CNS) involvement.[17] Similarly, it was found that patients with SLE performed significantly worse on both executive functioning and verbal fluency tasks.[18]

Given all this, I suspect that many, many lupus patients have cognitive deficits (and some even more moderate or severe cognitive impairment) without being diagnosed with CNS involvement and without being classified as NP+SLE patients. Most of us suffer in silence, either because we are embarrassed by our cognitive difficulties or because we are afraid that admitting them will take away our independence and our employment.

Neuropsychiatric Symptoms

Neuropsychiatric symptoms in lupus were first described by Moriz Kaposi in 1872 as "disturbed neurological function."[19] The ACR notes that 80% of people with SLE experience at least one neuropsychiatric symptom, **with cognitive impairment being the most common.**[20] Moreover, studies have also confirmed the high prevalence of cognitive impairment in lupus.[21] In fact, neurological involvement is almost a given at some point in the disease of lupus.[22] However, central nervous system involvement is less understood than other aspects of the disease.[23]

Moreover, a study assessed a group of patients with SLE to see if they had any of the 19 NP (neuropsychiatric) syndromes (as identified by the American College of Rheumatology 1999 committee report).[24] They found at least one NP syndrome in 91% of patients, with the most frequent deficit being **cognitive dysfunction**! Surprisingly, they dismissed these deficits (as well as headache and mood disorder) as being "minor" and "mild."[24] I can attest to the fact that MCI was **not** one of my milder symptoms and was not a **minor** issue in my life! It affects everything.

Functional Impact

Even when cognitive problems are milder, they can still have a significant functional impact on a person's life.[25] Some researchers divide neuropsychiatric syndromes or events into "focal" and "diffuse" events. A focal event would be something like a CVA/stroke, while a diffuse event would be a cognitive disorder or an anxiety disorder.[26] Researchers appear to agree that cognitive dysfunction falls into the diffuse category.[13] However, the mechanisms are still poorly understood. The individual factors behind diffuse damage are unclear but appear to involve inflammatory mediators.[19]

Research indicates that autoantibodies measured in the blood are more often associated with *acute* NP events, while autoantibodies measured in CSF are more often associated with *diffuse* events.[27] Other researchers have divided CNS lupus into thrombotic and nonthrombotic disease since different medications are more appropriate, depending on type.[28]

NP Positive or Negative?

I should mention that not all researchers believe that everyone with lupus has cognitive impairment, and a few studies have even found NP-SLE patients (those who haven't been diagnosed with neuropsychiatric lupus but have just plain lupus) to have normal cognition.

Moreover, a study completed neurological examinations and neuropsychological testing on SLE patients, which they divided into two groups: neuropsychiatric (NP+) and nonneuropsychiatric (NP-).[29] They found that the group of NP- patients performed at normal levels on cognitive tests, while the NP+ patients had difficulty with memory, psychomotor speech, and attention skills. This, however, is a little older study (2003), while several newer studies have identified mild cognitive deficits in NP- lupus patients, as previously mentioned. The confusion may lie in a lack of consistency in defining MCI or cognitive dysfunction(CD), what cognitive areas are tested, and how (what particular tasks or standardized tests or screeners were given).

Despite inconsistent findings in the literature, one thing is clear: the severity, course, & impact of cognitive impairment is greater in the lupus population than in the general population.[30]

Since multiple labels are utilized in the literature to describe less-than-optimal or below-average cognitive abilities, such as cognitively impaired or cognitive dysfunction or cognitive deficits, or even the more formally diagnosed MCI (mild cognitive impairment), I have combined all research studies using these various labels into this book. Since there is no consensus yet as to specific testing and labeling for these impairments in lupus patients, I thought it best to include everything related for a more complete picture.

NP Diagnosis

You may be wondering how a patient gets a diagnosis of neuropsychiatric lupus in the first place. It is my understanding that there was not a lot of consistency in the neuropsychiatric lupus diagnosis until the American College of Rheumatology (ACR) identified 19 different neurological and psychiatric syndromes of SLE in 1999. These indicate some type of nervous system involvement (think brain and/or spinal cord). In fact, 12 are considered CNS (central nervous system) syndromes, and 7 are considered PNS (peripheral nervous system) syndromes, but all are related to NPSLE (neuropsychiatric systemic lupus erythematosus).[31]

In addition, NPSLE can present in many ways: headaches, cognitive dysfunction, mood disorders, cerebrovascular accidents, transverse myelitis, neuropathy, etc.[32] As mentioned above, CNS syndromes are usually divided into focal (CVA) and diffuse (cognitive dysfunction) syndromes.[33] Cognitive dysfunction is only one of these 19 syndromes and is defined by the ACR.

Altered Physiology for All?

As mentioned, certain researchers have found greater cognitive impairment in NP+SLE patients (in areas of visuomotor coordination, attention, executive function, visual learning, memory, and phonetic fluency) as compared to SLE patients without other neuropsychiatric complications.[34] Relating this to neurophysiology, it has been found that NP+SLE patients have more pronounced atrophy of the bilateral hippocampus (smaller hippocampus) when compared to NP-SLE patients.[35] According to the researchers, the level of atrophy found suggested irreversible brain damage.

Many new studies have found subtle differences in lupus patients, even without other neuropsychiatric symptoms. A study "found no significant differences between the two patient groups (NP-SLE patients & NP+SLE patients), indicating that non-NPSLE patients, although they were clinically unaffected by neuropsychiatric symptoms, also displayed alteration in microstructural integrity (of the brain)."[36] Why would such individuals not yet show verifiable cognitive deficits upon testing? One theory is that certain people may have more cognitive reserve (spare cognitive capacity that protects the brain).[37] Another possibility is that these tiny brain changes show up before overt cognitive symptoms, as seen in the development of Alzheimer's disease.

In contrast to Zhang et al.'s study showing microstructural differences, a few studies have found that a third of SLE patients had normal MRIs.[38] Remember, though, that not all MRIs are at the same level of sensitivity, and other more sensitive imaging tests may be required to detect subtle differences. In fact, I have had two "normal" MRIs while experiencing definite cognitive impairment, so I think the damage would have to be more severe to show up on most traditional MRIs.

As you will see in the further research reviewed in this book, microstructural changes in the brain's white matter can be present long before anything shows up on an MRI. For review, white matter refers to the bundles of nerve fibers lying beneath the gray matter of the cortex, and even the water/fluid surrounding the white matter can be affected.[39] Researchers have also found alterations of white matter microstructure in **all lupus patients** (not just the NP+SLE subgroup) related to disease duration and fatigue.[40]

Thus, dividing lupus patients into NP+SLE and NP-SLE groups may not be beneficial when looking at the effects of lupus on the brain. Curiously, researchers have also failed to find any significant differentiating traits between NP+ and NP- SLE groups, such as race, age at diagnosis, disease duration, etc.[41]

Prevalence of Cognitive Impairment

Overall, even the prevalence of cognitive impairment in SLE patients is inconsistent across sources and studies. Most studies cite a wide range, such as 14-79% of SLE patients having cognitive deficits.[42] Another study states that cognitive impairment occurs in 60% of SLE patients.[43] However, other experts have reported that 75-80% of SLE patients experience cognitive dysfunction.[44] The highest estimate I found is that cognitive dysfunction may affect up to 90% of lupus patients.[45]

A study that specifically targeted SLE patients without CNS involvement (NP-SLE), found that *all* SLE subjects had cognitive impairment, which was statistically significant when compared to healthy control subjects.[46] Although pediatrics is not our focus here, one study of children with lupus found that 71% of them had cognitive dysfunction.[47] Similarly, another study found 65% of adult lupus patients to be cognitively impaired.[48] Finally, an additional study found that 58% of patients have cognitive impairment.[21]

How are we supposed to make sense of such a range and lack of agreement? Again, there is a high level of inconsistency across studies in the methods of each study, such as how they tested for cognitive impairment, how they defined cognitive impairment (i.e., which cognitive skills were measured), how they divided and characterized subjects, etc. Researchers say the wide range results from "non-standardized diagnostic criteria and screening tools."[49] It has also been pointed out that demographic factors and comorbidities may be reasons for the variation in prevalence rates.[42] So, at this point, no one knows how high the prevalence of cognitive impairment in lupus patients is, but it is clear that it is no small issue and one that requires further investigation. Indeed, it has been said that "cognitive dysfunction (in SLE) is a subtle (often missed) condition occurring with a high frequency."[50]

Lack of Consistency

Whether a difference exists between the two groups, we know that sometimes, in studies, NP+SLE and NP-SLE patients are separated, and sometimes they are not. Some have postulated that this discrepancy in research may help explain the wide range of prevalence rates. For example, a study tested SLE patients with a three-hour test battery and reported that almost 60% of patients had at least one cognitive function impairment, while 40% of patients had moderate or severe (not just mild) deficits.[51] A recent study found a 79% prevalence of cognitive dysfunction among SLE patients, but 80% of those patients had already been diagnosed with NP+SLE. An additional study found that 46% of regular SLE patients (NP-SLE) had cognitive impairment.[52]

Interestingly, a 2018 study found no significant differences in the number of white matter lesions between NP+SLE and regular NP-SLE patients.[35] Moreover, researchers found that **all** SLE patients (as compared to healthy controls) appear to have small deficits in certain areas (cognitive fluency and visual deficits), while NP+SLE patients tend to have "medium-sized deficits" in additional areas (i.e., attention, language), as well as larger deficits in the former areas.[53] They discuss that there is *a gradient of cognitive impairment*, with significantly greater cognitive impairment in NP+SLE patients.[53] It is also possible that many patients with SLE actually have NP+SLE, although undiagnosed. In fact, NP+SLE or "neurolupus" affects up to 75% of lupus patients.[54]

Another study has also speculated that education level, ESL (English as a second language) factors, medications, and anxiety/depression may also be reasons for the variability in the prevalence rates.[55] Neuropsychiatric disorders (depression, anxiety, mood disorders, prior stroke) are often associated with an increased prevalence of cognitive issues, as well.[42] Depression and anxiety may be tied to cognitive deficits, but are not the same and may have reversible components.[56]

General MCI Guidelines

After reviewing the research literature, the American Academy of Neurology (AAN) revised a guideline on MCI in the general population (not specific to people with autoimmune disease) in 2017. MCI prevalence in the general population varied by age, finding that 6.7% of people ages 60-64 had MCI, while 25.2% of individuals ages 80-84 had MCI.[57] MCI in the younger population (below 60) was not investigated. In this report, the AAN defined MCI as "a condition in which individuals demonstrate focal or multifocal cognitive impairment with minimal impairment of instrumental activities of daily living that does not cross the threshold for dementia diagnosis."[57]

Further, they divided MCI into two types: "amnesic MCI," in which memory difficulties are the predominant problem, and "nonamnesic MCI," in which other cognitive domains are impaired instead.[57] The distinction between these two MCI types is rarely seen in the SLE literature, but is used in the literature on Alzheimer's disease. General mild cognitive issues have been found in up to 78% of lupus patients.[25] In summary, considering the prevalence rates of cognitive deficits in SLE (although varied), it appears that these issues are much more common in patients with lupus than in the general population. This is certainly true for pediatric patients with lupus.

✶ ✶ ✶

In conclusion, cognitive impairment in lupus, often referred to as "brain fog," is a complex and multifaceted symptom that significantly impacts the lives of those affected. Despite its prevalence and severity, it remains underdiagnosed and poorly understood. The wide range of cognitive deficits, from difficulties with memory and attention to problems with word-finding and executive function, highlights the need for more comprehensive research and better diagnostic criteria.

Patients with lupus may experience a gradient of cognitive impairment, with varying degrees of severity that can fluctuate over time. This inconsistency in cognitive function can be incredibly frustrating and isolating, affecting daily activities, professional responsibilities, and overall quality of life. It is clear that more attention must be paid to understanding and addressing cognitive impairment in lupus patients, as it is a crucial aspect of the disease that requires targeted interventions and support.

As we continue to investigate the underlying mechanisms and potential treatments for cognitive impairment in lupus, it is essential to recognize the profound impact this symptom has on individuals' lives. By acknowledging and addressing these cognitive challenges, we can improve the quality of care and support for those living with lupus, ultimately enhancing their overall well-being and quality of life.

Chapter 5

How Does Cognitive Impairment Appear in Lupus?

Cognitive impairment in systemic lupus erythematosus (SLE) is a deeply personal and often misunderstood symptom—one that remains insufficiently explained despite decades of research. While lupus is typically associated with joint pain, fatigue, and organ involvement, the way it quietly erodes mental clarity, memory, and word retrieval is rarely discussed with the same urgency. For many patients, the effects on thinking and memory are not just frustrating; they're isolating and frightening.

This chapter explores the cognitive challenges individuals with SLE face, beginning with my experience of unexpected lapses that disrupted my professional and personal life. Through a blend of lived experience and clinical findings, we will examine how cognitive dysfunction shows up in lupus patients, what science currently understands about its origins, and how this invisible symptom affects daily functioning and emotional well-being. Whether you're navigating these struggles yourself or trying to understand a loved one's journey better, this chapter offers validation, insight, and clarity on a topic that deserves much more attention.

More Than Fatigue: The Cognitive Toll of Lupus

Cognition in SLE remains incompletely understood.[1] For me, it all started with word-finding. I was teaching a graduate-level college course and couldn't think of a professional word I used frequently. It didn't happen just once, but it began happening more and more frequently, both in the classroom and at home. Then, memory issues appeared, as one ordinary day, I went to the grocery store, picked up the kids from school, and completely forgot to unload the groceries. By the time I remembered, the cold stuff was ruined.

In another instance, someone asked me what my little girl was for Halloween (two weeks prior), and I couldn't remember. I told them about last year's costume so as not to look like an idiot. I remembered a few minutes later, but the moment had passed. How could I forget the costume I had ordered and put on her three different days? Each of these instances was so embarrassing to me. I reveal them here because if you have experienced something similar, I want you to know that you're not alone. It's not your fault—it's autoimmune cognitive impairment.

These may not be life-threatening mistakes, but they can definitely be life-altering, as I began to worry about my ability to work and care for my family every day. Speaking of work, while teaching a graduate level speech-language pathology class one day, a student told me she was working with a woman with lupus on her cognition. I remember screaming in my mind, "That was me!" What I said out loud was that, yes, autoimmune diseases can definitely affect cognition, and I was glad her client was getting treatment. On days when my own cognitive impairment was more significant, I would have to ask my husband to proof a work e-mail prior to sending it. I felt I could not trust my own thinking, which was very disturbing, and something I hear often from clients I serve. Do you see now why impaired cognition was my most hated lupus symptom? Not only could I not count on my body anymore, I could no longer count on my brain!

How Does Cognitive Impairment Manifest?

Recently, a study stated that "cognitive dysfunction is an insidious and underdiagnosed manifestation of systemic lupus erythematosus that has a considerable impact on quality of life, which can be devastating."[2] Furthermore, a group of researchers revealed that "patients often report CI (cognitive impairment) as the most bothersome disease-related manifestation, with a great effect on their quality of life."[3]

In 2018, Lampner interviewed Dr. Zahi Touma from the *Centre for Prognosis Studies in the Rheumatic Diseases*.[4] Dr. Touma said that patients with SLE report specific deficits in attention, recall, concentration, information processing, spatial processing, and word-finding, and he discussed how these cognitive problems can negatively affect quality of life. I experienced all of these types of deficits in the past. I already mentioned that I first noticed word-finding (anomic) deficits, but I also frequently experienced difficulty with attention (sustained) and concentration on a topic, as, for example, when I was writing a lecture or grading a paper. I also had difficulties recalling information after listening to a speaker and reading, so I compensated by taking many more notes.

Information processing was also an issue, as, at times, I have had difficulty considering new information, assimilating it into my previous knowledge, and arriving at a solution or conclusion that would have been effortless for me before lupus affected my brain. When processing or trying to recall information (or specific words), it was as if I knew it was in my brain, but I couldn't find the right pathway to get to it. That was extremely frustrating!

ACR Criteria

The ACR (American College of Rheumatology) identified eight areas of cognition often affected in lupus: "simple attention, complex attention, memory, visual-spatial processing, language, reasoning/problem-solving, psychomotor speed, and executive functions, at least one of which must be affected in order to meet the case definition of cognitive dysfunction."[5]

In a systematic review of the literature, it has been concluded that the most common cognitive deficits lupus patients experience are related to verbal and nonverbal learning and working memory.[6] Less common deficits include attention, psychomotor speed, and executive function deficits (planning, organization, etc.).

Patients with SLE have been found to have poorer working memory performance (difficulty encoding and manipulating new information) than normal healthy control subjects.[7] As you probably know, working memory is also referred to as short-term memory and is thought to be controlled by the "central executive system of the brain."[8] As working memory demands increase, SLE patients have slower reaction times, demonstrating difficulty with processing speed. The researchers concluded that lupus patients struggle with working memory and processing speed as separate deficits, which were evident in both NP+ (neuropsychiatric) SLE patients and NP-SLE patients.[7] Moreover, a group of renowned researchers found lexical and semantic memory deficits in lupus patients, although spatial working memory was affected the most.[9] Another study tested 58 lupus patients and found that the visuospatial domain was the most affected, which they said suggested a fronto-parietal deficit.[10]

When cognitive deficits first appear, it is possible that these would be apparent to only the patient when, for example, a person who would normally score above average on cognitive tests now only scores in the average range. Professionals who are scoring cognitive assessments would see the person as "within normal limits" or "normal" when, in fact, there had been a definite decrease in performance. In a similar vein, research pointed out that a person's "average" test scores might "represent a significant decline in their cognitive and functional ability."[5] This may partially explain what happened in the following study when scientists found lupus patients to have "lower average" range scores. Cognitive flexibility, processing speed, and attention were all found to be in the lower average range for SLE patients when compared to matched peers, regardless of the age of the patients.[11]

One study found psychomotor speed (as measured by a Digit Symbol Substitutions test) to be the most significant deficit in SLE patients who had not been previously diagnosed with cognitive impairment.[12] Other researchers have observed impairment in objective performance and self-reported functioning across cognitive and physical domains.[11]

When Does it Appear?

There is controversy regarding when cognitive dysfunction appears in the progression of lupus and whether it is directly related to damage and flares in lupus. Is it related to a more severe disease process, or does it tend to show up in people who also have more systemic organ damage because of lupus? Does it show up only when a patient is "in a flare"? SLICC-DI (Systemic Lupus International Collaborating Clinic Classification Damage Index) is a common measure of lupus-related damage. Using this measure, researchers found cognitive dysfunction was unrelated to disease activity in their study.[13] However, another study found that chronic disease damage (no matter what organs were involved) is indeed linked to cognitive impairment.[14]

Moreover, a positive correlation between disease duration and increasing cognitive impairment was found.[15] Surprisingly, a study found SLE patients in a flare did better on cognitive testing than those with inactive disease.[16] Perhaps, then, current disease activity (or whether or not the patient is in a flare) may not matter as much as chronic systemic damage over time. Similarly, it was also found that cognitive impairment was associated with disease damage, also as measured by the SLICC-DI.[17] This makes sense to me, such that as your disease progresses and you accumulate more and more damage to your body, your brain would more likely be negatively affected, as well.

When I consider my own story, I don't think that I had cognitive issues until approximately ten years after my first lupus symptoms. I wondered if this was typical for SLE patients or if some patients experience cognitive struggles at the very beginning of the disease. I think it is also important to recognize that "there is a long asymptomatic phase between the onset of neuropathological processes and the onset of cognitive impairment in cognitive disorders."[18] However, it was concluded in a study that "the relationship between cognitive dysfunction and overall disease activity in SLE remain unclear."[5]

Patient Recognition of Impairment

Cognitive dysfunction has been reported to affect anywhere from 6-66% of people with lupus, "but the prevalence may be up to 95% when cognitive defects are assessed by computerized neuropsychological testing."[19] Another significant question is whether patients always recognize their own cognitive deficits. I think I was sensitive to my mild cognitive issues because of my profession and clinical experience with cognitive deficits, as well as the fact that I could see it immediately affecting my work and home life. Perhaps some patients laugh it off as aging or being scatterbrained, or maybe they don't really notice a change in their daily lives, at least at first.

In a study, 21 patients had abnormal cognitive scores.[14] Six of these patients self-reported definite thinking and memory problems, and nine indicated occasional thinking and memory problems, but seven of the affected patients denied any cognitive problems. I wonder whether those seven patients truly had not noticed any difficulties with their thinking/memory or if they were embarrassed or afraid to admit it for some reason. Certainly, I can relate to that embarrassment, even among family and friends, not to mention the very real fear of losing your job and income as a result of cognitive deficits. We're talking about serious life alterations here.

Remember that deficits are always individualized, of course. As Harrison pointed out, one single thought process may be affected at a time or many thought processes at once.[20] Also, recall that some type of cognitive impairment is often present in SLE, even in patients without documented or obvious signs of central nervous system involvement.[21] Generally speaking, researchers stated, "cognitive problems experienced by individuals with MCI negatively impact their lives, including mood, relationships, treatment compliance, and independence."[22]

Confusing Terminology

Throughout this book, you will notice various terms utilized when referring to cognitive deficits, such as cognitive impairment or cognitive dysfunction. The intent here is not to be confusing. Different journal articles and books utilize alternate terms. I have tried to be true to the terminology used by each source. When cognitive impairment or cognitive dysfunction are used, consider them equivalent terms, as they refer to a cognitive deficit of some kind. Sometimes, authors or researchers put qualifiers before the terms, such as "severe cognitive impairment" or "mild cognitive dysfunction."

Moreover, there is a generally accepted continuum to cognitive impairment, as described by Dr. Bredesen (in his work on Alzheimer's progression) in which subjective cognitive impairment (SCI) is more subtle than MCI, which is less significant than dementia.[23] Alzheimer's is often considered the most severe form of dementia, although there are other severe forms of dementia, such as Lewy body dementia. Dr. Bredesen teaches that people may have SCI (with changes only the individual notices) up to ten years prior to developing a diagnosable cognitive impairment.[24] Other researchers, doctors, and scientists tend to ignore and dismiss SCI and even MCI and don't appear to "count" cognitive deficits until they reach significant dementia proportions. Dr. Bredesen's work is discussed in more detail later in this book, but I have been so impressed with his results that I pursued further education and am now a certified ReCODE 2.0 health coach (which means I am certified to coach the program he developed for prevention and reversal of cognitive decline).

In addition to levels of impairment, I believe the lack of therapeutic treatment and prevention options does a great disservice to the patient, as there have been documented physiological brain changes even before SCI develops. Further, remediation programs have been proven to be more successful and efficient when the patient has milder deficits, so it is important to identify them early. This is why I am so disturbed by certain studies with autoimmune patients, in which researchers sometimes dismiss brain fog or MCI as something less than a real neuropsychiatric event. Waiting to see if it progresses before you do something about it is a very poor choice with significant life consequences.

✶ ✶ ✶

Cognitive impairment in systemic lupus erythematosus is a complex, life-altering symptom that deserves far more recognition than it often receives. Throughout this chapter, we have explored how cognitive deficits manifest, how they may progress over time, and how deeply they impact daily functioning, relationships, and emotional well-being. Whether it starts with word-finding difficulty, memory lapses, or struggles with attention and information processing, the experience can be deeply isolating and frightening. Importantly, research confirms that cognitive dysfunction in lupus is both common and underdiagnosed, often appearing independently of obvious disease flares or active inflammation. This knowledge provides critical validation for those living with these challenges.

Understanding the full spectrum of cognitive symptoms, their potential causes, and their real-life consequences allows patients, families, and healthcare providers to approach lupus care more holistically. By sharing my personal experiences alongside clinical findings, my hope is to encourage greater compassion and urgency around this often invisible symptom. Cognitive struggle in lupus is not a reflection of a person's abilities or efforts; it is a manifestation of a complex disease process. Recognizing and addressing cognitive impairment is a vital part of supporting quality of life for those living with SLE.

Part II:

Investigating the Causes

Chapter 6

Where is the Problem in the Brain?

Cognitive impairment in systemic lupus erythematosus (SLE) is a complex and often frustrating phenomenon, both for patients and researchers. Despite growing awareness of the neurological symptoms associated with SLE, the precise brain regions and physiological mechanisms responsible for cognitive decline remain poorly understood. In this chapter, I review the neurophysiology underlying cognitive dysfunction in SLE, drawing on emerging research from neuroimaging studies and functional MRI data. This includes changes in white and gray matter, brain volume abnormalities, altered functional connectivity, and compensatory mechanisms in various brain regions. Rather than pointing to a single culprit, findings suggest a network-based disruption involving multiple areas and processes. As we continue to uncover new aspects of brain anatomy and function, our understanding of how SLE affects cognition evolves—highlighting both the challenges in pinpointing the exact origin of impairment and the importance of continued interdisciplinary research.

Understanding Cognition and Its Corresponding Physiology

Researchers define cognition as "the sum of intellectual functions that result in thought."[1] On a physiological level, Mackay defines cognition as "the functional result of a process whereby learning and memory result from communication between neurons, astrocytes, glial cells, and immune cells through a variety of neurotransmitters, transcription factors, cytokines, and chemokines."[2] She goes on to point out that the "hippocampus is the primary brain structure involved in memory and cognition."[2]

Despite this understanding, the "neural substrates" of cognitive deficits in SLE are unknown, making alleviating symptoms challenging until we can understand more of what is truly happening.[3] As Zhang and his colleagues stated, "understanding the pathophysiologic mechanisms leading to NP+SLE is essential for the evaluation and design of effective interventions."[4]

MRI Studies

MRI and functional MRI have certainly improved our understanding of the neurophysiology of cognitive impairment and offered some possible explanations, although we are far from a consensus. Several studies have pointed to specific changes in white matter (nerve fibers/axons) to explain cognitive difficulties in SLE patients. For example, poor complex attention has been linked with white matter tract abnormalities, and similarly, slower processing speed was linked with decreased white-matter integrity.[5] White matter hyperintensities have been associated with cognitive dysfunction in general.[6]

Further reinforcing this connection, studies found SLE patients to have significantly more white matter lesions on MRI when compared to healthy controls and correlated the lesions to cognitive dysfunction, particularly verbal memory.[7] One such study concluded that "patients with SLE have more microstructural brain white matter damage" than controls.[8] Indeed, they found that mean diffusivity (tracking how diffuse or spread out water molecules are along white matter bundles) was higher in eight major white matter tracts in SLE patients than in healthy controls.[3]

Similarly, a systematic review found that most studies that utilized diffusion tensor MRI found significantly increased mean diffusivity values in several white matter regions of all SLE patients (whether NP+SLE or not) compared to healthy control subjects.[9] In addition, white matter hyperintensities, while also found in normal controls, are significantly higher in patients with SLE and may contribute to disruptions in network connectivity.[10]

However, researchers caution that "diffuse white matter damage" may not always appear on conventional MRI testing.[11] Indeed, over half of NP+SLE patients have "normal MRIs," as they are mostly sensitive to focal findings rather than white or gray matter lesions or atrophy.[12] Similarly, another study found that up to 45% of lupus patients with verifiable neuropsychiatric symptoms had no visible abnormalities on MRI.[13] I have previously shared that this was true in my case, as I had two "normal MRIs," while experiencing cognitive deficits. MRIs vary by sensitivity and accuracy, and they do not detect all changes.

Decreased Brain Volume & Abnormalities

Many studies have found atrophy (decreased volume) in various brain areas tied to MCI. Some researchers think such brain atrophy is associated with additional SLE-related organ damage.[14] Furthermore, a study utilizing MRI scans found cerebral and corpus callosum volumes to be significantly smaller in patients with SLE (compared to healthy controls).[15] Moreover, such atrophy was correlated with cognitive impairment.[15]

Specifically, grey matter volume was found to be reduced in both the insula and parahippocampal regions of SLE patients, while VMHC (voxel-mirrored homotopic connectivity comparing right and left hemispheres) values were increased in these same areas and were positively correlated with decreased attention and abstraction abilities.[16] White matter atrophy of the left frontal lobe has also been specifically linked with poorer working memory in SLE patients.[17] Researchers cite the following cortical regions as susceptible to damage in SLE: hippocampus, parahippocampus, putamen, and orbitofrontal/prefrontal cortex.[18]

A recent study posed that a decline in synergy between the putamen/basal ganglia and cerebellum (vermis 6) caused cognitive dysfunction in SLE patients.[19] Using fMRI, it was found that SLE patients without other neuropsychiatric symptoms had a "subclinical cognitive dysfunction and decision-making deficit," and they correlated this with lower activation in two prefrontal regions: the ventromedial prefrontal cortex and orbitofrontal cortex.[20] Similarly, researchers found that SLE patients accumulated damage to both the hippocampus and the amygdala over time.[21] Another study found only two of nine brain regions to be abnormal (cerebellum and hippocampus), although they were measuring the distribution of a mitochondrial translocator protein (TSPO) within the brain.[22] They concluded that the altered distribution of the TSPO protein may be a marker of pathophysiology in the brain.[22]

By utilizing specialized MRI scans, another study found a reduction in multiple metabolite concentrations in the posterior cingulate gyrus of lupus patients, both those with and without neuropsychiatric symptoms.[23] Interestingly, a recent study observed that the cerebrospinal fluid (CSF) from lupus patients was neurotoxic (toxic to the nervous system).[24]

Compensatory Brain Regions

Other researchers have discovered larger than normal cortical areas in lupus patients, indicating "recruitment" of other areas to compensate for difficulty with working memory tasks.[25] In this situation, patients use more or larger areas of the brain in order to compensate for possibly "damaged" areas. It was concluded that SLE patients use compensatory mechanisms in the frontal and parietal regions to help them with cognitive tasks such as working memory, attention, and executive function tasks.[26] Using fMRI and MRI, researchers found SLE patients had increased bilateral cortical activation in the frontal lobes during executive function and working memory tasks, which demonstrated cortical overactivation as compensatory for early white matter neuropathology.[25]

In fact, evidence for compensatory mechanisms was also found in children with lupus who utilized fMRI during cognitive tasks. These children could use compensatory brain regions for attention, but these compensatory mechanisms broke down for visuospatial and working memory tasks.[27] Corroborating this, fMRI studies have verified that portions of the frontal and parietal regions are process-general or active during a wide range of demanding cognitive tasks.[28] Thus, the same voxels (specific brain volume areas) light up when inhibiting and storing information for working memory tasks and solving math problems. These functionally-general areas help humans have cognitive flexibility. Therefore, SLE patients may need to tap into these general regions more often if other areas are damaged or neural connections are interrupted in domain-specific areas.

A similar explanation is found in "less wiring, more firing," as increased cortical activity is attempted to overcome early white matter changes in cognitive impairment.[25] Related specifically to lupus patients with cognitive deficits, researchers recently found evidence for a compensatory mechanism in that hypermetabolism in the anterior putamen and frontal cortex indicated the recruitment of additional brain regions to successfully complete a spatial navigation task.[29] Moreover, some researchers found hypermetabolism tied to memory regions like the hippocampus, the orbitofrontal cortex, and basal ganglia in SLE patients.[30] It may be that brain changes initially allow SLE patients to utilize compensatory mechanisms with increased activation in certain regions, but these mechanisms eventually fail, which is when cognitive deficits become more obvious.[27]

Abnormal Networks

The fact is that linking brain regions to specific cognitive functions can be difficult. It used to be thought that a single brain area correlated to a single thought process. Newer cognitive neuroscience theories, however, discuss "complex cross-modal interactions where conjoint functions of brain areas work together in large-scale networks."[31] Studies found abnormal brain networks in SLE by utilizing graph theory for such nodal networks, along with functional MRI (fMRI).[19]

Specifically, SLE patients had decreased network functional connectivity between the basal ganglia and the cerebellum and decreased efficiency in the right insula, bilateral putamen, and bilateral Heschl's gyrus.[19] In such models, cognitive networks are formed in which several brain regions communicate to accomplish a single task. A specific cognitive task is no longer thought to be linked to one specific anatomical brain region alone, so we cannot pinpoint one region that is negatively affected in lupus patients with cognitive issues. We are still discovering brain anatomy and physiology, such as a new layer of tissue around the brain called the subarachnoid lymphatic-like membrane. Therefore, we have not yet found clear, specific connections between cognitive impairment and autoimmune diseases.

In conclusion, cognitive impairment in systemic lupus erythematosus is a multifaceted issue that does not arise from a single brain region or mechanism. Instead, it appears to reflect widespread disruptions across brain networks involving white matter integrity, gray matter volume, and functional connectivity. While MRI and functional MRI studies have offered valuable insights, many patients with neuropsychiatric symptoms continue to present normal imaging results, highlighting the complexity of detecting subtle neurophysiological changes. The evidence for both compensatory brain mechanisms increasing cortical activation and abnormal network connectivity further underscores the brain's attempt to adapt to early and ongoing damage.

However, these compensatory efforts may only be temporary, eventually giving way to noticeable cognitive deficits. The evolving understanding of brain anatomy, including new discoveries like the subarachnoid lymphatic-like membrane, reminds us that our overall knowledge of the brain remains incomplete. As research progresses, it is becoming clear that cognitive dysfunction in lupus cannot be reduced to a single pathology. Instead, it reflects an intricate interplay of immune, vascular, and neural factors across multiple systems. Future research must continue to embrace interdisciplinary approaches in order to better identify early markers of cognitive decline and design more effective interventions for patients living with the cognitive challenges of autoimmune disease.

Chapter 7

What Causes These Issues?

Understanding the underlying causes of cognitive impairment in neuropsychiatric systemic lupus erythematosus (NP+SLE) remains a complex challenge. While much of the pathogenesis is still being debated, various hypotheses have emerged, shedding light on the potential mechanisms contributing to cognitive dysfunction. These include systemic inflammation, neuronal injury due to autoantibodies, cerebrovascular damage, medication effects, and chronic disease-related damage. In addition, factors like vasculopathy, antibody-mediated neurotoxicity, and cytokine signaling abnormalities may play significant roles.

This chapter will share these multifactorial theories, reviewing the complex interactions between the immune system, neurotransmitter imbalances, neurogenesis, and objective biomarkers. These emerging insights highlight the intricacies of cognitive decline in lupus and emphasize the need for ongoing research to understand better and address the cognitive challenges faced by individuals with this autoimmune disease.

Understanding Cognitive Impairment in Lupus

As we are well aware by now, "Diagnosing and treating neuropsychiatric systemic lupus erythematosus remains challenging as the pathogenesis is still being debated."[1] Nevertheless, we are beginning to see that there are many hypotheses regarding the pathophysiology of cognitive impairment in SLE. These include, but are not limited to, overall inflammation, neuronal injury due to antibodies, cerebrovascular atherosclerosis, small vessel damage, medication effects, and chronic disease damage.[2] Similarly, Lampner states that "vasculopathy, antibody-mediated neurotoxicity, impairments in cytokine signaling, and/or complement abnormalities" could be contributing factors to cognitive deficits in SLE.[3] Remember that the reference to "contributing factors" might mean multiple causes or multiple contributors to the cognitive impairment found in lupus.

Similarly, a study concluded that "the pathogenesis of NP+SLE is multifactorial and involves various inflammatory cytokines, autoantibodies, and immune complexes resulting in vasculopathic, cytotoxic and autoantibody-mediated neuronal injury."[4] In addition, "lupus related autoantibodies most frequently associated with NP+SLE include antiphospholipid antibodies, anti-ribosomal P antibodies, and autoantibodies, which bind to neuronal antigens such as the N-methyl-D-aspartate receptor."[5]

Moreover, it has also been revealed that "the possibility of specific antibody-induced brain injury in SLE remains an intriguing possibility."[6] In addition, it has also been stated that "cognitive dysfunction in SLE sits at the interface of two of the most complicated organ systems in the human body, namely, the nervous system and the immune system."[7] I'll outline various factors that may contribute to cognitive decline in autoimmune disease.

Systemic Inflammation

High levels of systemic inflammation have been frequently associated with cognitive impairments. A recent study states that "systemic inflammation associated with chronic rheumatological diseases has been postulated to be a key driver of cognitive decline."[8] Indeed, systemic inflammation is a known driver of cognitive decline in patients with chronic diseases and the general geriatric population.[9] This connection also held true in one study of lupus patients, in which circulating IL-6 (Interleukin-6, an important cytokine) levels correlated inversely with performance on memory testing.[10]

Expanding on this inflammatory connection, a recent study of mice with lupus showed that LCN2 (lipocalin-2) activates astrocytes, which promote pathological neuroinflammation.[11] Pro-inflammatory cytokines (like IL-6 and TNFalpha) are often associated with brain-destructive processes (including cognition and fatigue), but research has also shown them to be of critical importance in normal synaptic function and hippocampal memory.[12] It seems to be an issue of **balance** with cytokines. Indeed, Mackay says that cytokines have variable effects on learning and memory, so the magnitude and duration of cytokine exposure are important.[12]

To further support the inflammation hypothesis, the research found cerebral atrophy in the brains of SLE patients and hypothesized that it was due to "an inflammatory process, cytokines, or locally produced autoantibodies."[13] This relates to overall inflammation levels in the body, as we must have a certain amount of inflammation to fight infection and heal ourselves. Still, too much (and chronic) inflammation can lead to multiple complications and diseases.

Inflammation and CRP

Also related to inflammation, another study found that SLE patients with high C-reactive protein (CRP) levels (a blood test measure of systemic inflammation) had cognitive deficits, particularly working memory difficulties.[14] Researchers concluded that this suggests "inflammatory involvement of brain parenchyma." High C-reactive protein levels have been associated with increased microglial activation, which produces inflammatory cytokines and can lead to the death of neurons.[9]

I have also struggled with high CRP levels while dealing with cognitive deficits. My CRP levels were too high (which also places a person at higher risk for heart attack, stroke, and cancer), then decreased after several months of Benlysta (the first medication ever specifically created and approved for lupus treatment). Interestingly, my cognitive symptoms also decreased around the same time. To my disappointment, however, my CRP levels began to slowly rise again about a year later (while still receiving Benlysta infusions every four weeks), while my cognitive deficits increased again. Again, the rise in CRP level indicates increased systemic inflammation somewhere in the body.

It is interesting that systemic inflammation and cognitive impairment have been linked in at least two studies.[10,14] It has also been suggested that inflammation may occur when B-cells and autoantibodies "traverse the blood brain barrier, promoting an inflammatory environment consisting of glia activation, neurodegeneration, and consequent averse behavioral outcomes."[15] Inflammation is probably not the whole picture. However, as Boumpas suggests, neuroinflammation is only one of three key mechanisms, with the other two being cerebral ischemia and blood-brain barrier disruption.[16]

Neurotransmitter Levels

Another factor to consider is whether neurotransmitter levels, such as norepinephrine and serotonin, affect cognition. Long-term potentiation (LTP) is the physiologic process by which the hippocampus lays down new memories.[12] It has been shown that low levels of neurotransmitters help the LTP process, but high levels are disruptive.[17] Yet another mechanism that can negatively affect hippocampus performance includes anti-NMDAR and anti-P antibodies, which mediate cognitive disturbances in the hippocampus and amygdala, as concentrations should not be too low or too high.[12] Again, as in systemic inflammation, we see that a certain **balance** is needed, in which levels are neither too low nor too high.

Neurogenesis

Another possible theory has to do with adult neurogenesis (making new neurons). Neurogenesis occurs in the lateral subventricular zone and the dentate gyrus of the hippocampus.[12] It has been shown to contribute to higher hippocampal function, which helps learning and memory.[18] Theories of "neuronal degradation" are supported by certain studies.[19] Animal models show that inflammatory processes in the brain may lead to neuronal loss and impaired neurogenesis.[20]

Moreover, lupus patients generate autoantibodies, which may cross-react with other antigens and cause compromise of the blood brain barrier, leading to neuronal death and subsequent cognitive dysfunction.[21] Autoantibodies may specifically injure and target neural cells.[22] In addition, stress and inflammation both have a negative effect on adult neurogenesis.[23] This may impair performance on cognitive tasks. There is possibly a link, then, between too much systemic inflammation and a lack of adult neurogenesis.

Objective Biomarkers

Scientists have sought out objective measures taken from blood or CSF (cerebrospinal fluid) samples, which may correlate with and identify cognitive impairment in lupus patients. For example, S100B protein levels were found to be increased in NP+SLE patients with cognitive dysfunction.[24] These researchers stated that the S100B protein plays a crucial role in stimulating and inhibiting inflammatory activity. Higher S100B levels have been linked to brain injury and permeability of the blood-brain barrier.[24] They are also associated with exacerbation of neuronal death and decreased neuronal function.

In contrast, a different study suggests that taking a sample of CSF of lupus patients may result in a titer of AAb, which indicates the severity of cognitive dysfunction.[21] The researchers hypothesize that this result implies neuronal death and/or synaptic dysfunction.[21] A more recent study concluded that there is "potential to utilize CSF immune transcriptome changes to identify disease-associated neuroinflammation" in those who are cognitively impaired.[25]

Additionally, a group of researchers found NP+SLE patients to be more likely to have various elevated autoantibody serum levels as compared to SLE patients without neuropsychological impairments.[26] When both serum and CSF were tested in SLE patients, some with neuropsychiatric issues and some without, it was found that those with neuropsychiatric symptoms had significantly higher levels of neurofilament light chain and concluded that this was a marker of neuronal damage.[27]

If future studies corroborate that such objective samples are linked to cognitive dysfunction, researchers would then need to confirm whether these levels were involved in the pathology of these deficits or were simply indicators. In an alternate CSF theory, it has been stated that evidence supports the hypothesis that brain reactive antibodies (BRAs) taken from the CSF of lupus patients can bind multiple antigens, induce neuronal apoptosis (death), and, thus, alter behavior.[28]

Although CSF is supposed to provide immune protection for your brain, a new study showed that in cognitively impaired patients, inflamed T-cells were entering the brain due to a certain cell receptor (CXCR6) receiving signals from degenerating microglia.[25] Non-neuronal cells, such as microglia, oligodendrocytes, and astrocytes, have been found to experience more significant changes during aging than neurons.[29]

Vascular Disease

A study sought to determine if cognitive dysfunction in lupus results from persistent inflammation (whether measured as ongoing disease, high autoantibodies, cytokines/chemokines) or from chronic damage, critical medical events, and/or comorbidities.[30] They found no difference in disease activity between the 16% of lupus patients that had cognitive impairment and the ones who did not. They concluded that their results did not support a definitive inflammatory mechanism for cognitive deficits. Still, they suggest cognitive impairment may arise from chronic damage, which is probably of vascular (blood vessels) origin.

Related to the vascular theory, a study found an increase in cerebral small vessel disease in a small sample of SLE patients, which, they concluded, caused cognitive impairment symptoms.[31] SLE patients are at a higher risk of atherosclerosis (lesions in artery walls that lead to narrowing), and one study found that SLE patients with cognitive problems had higher levels of IgG antibodies.[30]

Other researchers have stated that, at least for NP+SLE, cognitive impairment is due to vascular injury arising from the autoimmune response.[32] Research indicates that focal manifestations may be due to vascular damage, while diffuse manifestations in SLE may be immune-mediated.[33] It has been hypothesized that SLE deficits result from both inflammation (due to autoantibody damage to tissues) and vasculopathy.[34]

Vascular changes may happen at a level undetectable by MRI, as sections of cerebral vessels demonstrating micro-thrombi with C4d and C5b-9 deposits have been linked to NP+SLE.[35] Along these lines, another study found evidence of microvascular degradation and neuroinflammation in SLE patients who did not have active disease by finding an association between reduced cognition and white matter extracellular free water (water surrounding white matter) increases.[36]

Axonal Damage

Another theory is that damage to myelin and axons may be the cause of a central nervous system disturbance, which results in cognitive dysfunction.[22] Such a demyelinating process would be similar to what is thought to happen in MS (Multiple Sclerosis). It has been proposed that this type of axonal damage or demyelination could lead to reduced white matter volume in lupus patients, while decreased gray matter volume may be caused by cortical atrophy.[37]

Meningeal inflammation (of protective membranes covering the spinal cord and brain) may be another possibility.[22] It is possible that a marker of axonal loss—N-acetylaspartic acid (NAA)—may be tracked in future studies of SLE patients.[22] It can help to investigate an association between axonal loss and MCI.

Corpus Callosum

One study showed atrophy of the corpus callosum to be associated with cognitive impairment in lupus patients.[13] Although, to my knowledge, these findings have not been replicated. Moreover, a study found decreased white matter integrity in the corpus callosum (as well as in the anterior corona radiate) of NP+SLE patients.[38]

In a case study of one patient with NP+SLE, researchers found vasogenic edema (swelling) in the corpus callosum, which decreased with steroid pulse therapy.[39] However, they found no differences between SLE patients and healthy controls in corpus callosum size overall.[40]

Blood-Brain Barrier

As previously mentioned, others have hypothesized that the blood-brain barrier (BBB) has somehow become permeable in lupus, and antibodies may cross the barrier and damage the brain. The BBB is essential for neuronal function and cerebral homeostasis.[41] In fact, the BBB has frequently been "proposed as the entry point of systemic effectors that cause inflammation and injury."[1] A group of well-known researchers found that patients with lupus had significantly higher levels of BBB leakage than normal controls and that this leakage was linked with changes in brain structure (smaller gray matter volume) and cognitive impairment.[42] An association between leakage of the BBB and neural pathology underlies cognitive impairment in patients with lupus.[43] These neural changes, then, cause different patterns and a decreased number of connections between the brain regions of these individuals. Extensive BBB leakage was found by contrast-enhanced MRI to be directly related to cognitive impairment (assessed by formal neuropsychological testing) in 75% of cases, while cognitive impairment was not associated with the presence of autoantibodies.[44] Dr. O'Bryan calls this B4: Breach of the Blood Brain Barrier.[45]

Now, how does the BBB become affected? Theories include entry through the blood supply, the choroid plexus, the glymphatic system, or the meningeal-arachnoid barrier.[46] Antiribosomal P and anti-NR2 antibodies, along with complement activation, are another theory of the cause of BBB leakage.[42] Furthermore, researchers caution against favoring the BBB as a mechanism for CNS disease in lupus while ignoring other options, such as the meningeal barriers and cerebrospinal fluid barrier.[47] In fact, in a mouse model, studies found no evidence of BBB dysfunction; instead, they found a breakdown in the blood-cerebrospinal fluid barrier (BCSFB), which, they say, is what allows antibodies to reach the brain and lymphocytes to enter the ventricles.[48] They conclude that dysfunction with the BCSFB barrier is a causative factor in neuropsychiatric lupus, which also includes cognitive impairment.

Other researchers agree that recent experiments with mouse models show that BCSFB leakage may be responsible for letting proinflammatory molecules near the brain instead of directly passing through the BBB.[49] Some researchers postulate that BBB damage may be present but develops slower than damage to other organs. If the BBB breaks down, then antineuronal antibodies may be allowed into the brain parenchyma.[33] The compromise of the BBB may acquire damage over time, as the BBB provides some protection but breaks down with repeated exposure to antibodies, cytokines, and activated immune cells.[50] Disruption of the BBB may be caused by improper regulation of the specific cytokines IFN-alpha, tumor necrosis factor (TNF), or IL-6.[51] IFN-alpha is the "cytokine with the best described relationship with cognitive dysfunction."[46] Some researchers hypothesize that the BBB is compromised in lupus patients, allowing autoantibodies into the brain and causing cognitive difficulties.[52] Researchers state that there is "growing evidence supporting the presence of BBB dysfunction in NP+SLE."[53]

Furthermore, environmental factors (stress, ischemia, infections), coupled with inflammatory cytokines, can damage the brain in various areas, contributing to different symptoms.[54] For example, it has been found that IL-6 (interleukin 6) and C3 (complement component 3) were increased in an "acute confusional state" due to their entry from CSF through the damaged BBB.[55] A combination of factors, such as complement peripheral cytokines, autoantibodies, and factors unrelated to autoimmune disease, may all compromise the BBB in lupus patients.[56]

Central Nervous System Involvement

A group of researchers found cognitive impairment to be associated with a longer duration of central nervous system involvement.[57] More generally, studies found that longer disease duration is associated with poorer cognitive function in SLE patients.[58] It was thought that both current disease activity and chronic damage influenced cognitive impairment.[59] They even hypothesized that preventing disease flares and relapses could control cognitive deficits.[59] Research has also found disease duration to be related to reduced cerebral and corpus callosum volumes.[59]

Researchers said their findings suggest that chronic disease damage is linked to cognitive dysfunction in lupus patients.[2] Perhaps it is, in fact, all about inflammation and that the "end result of inflammatory damage to white matter can be cognitive dysfunction," but it is a "process you can come to by several different mechanisms."[60]

Summing up the various theories of pathophysiology, researchers recognized two main pathways as an ischemic-thrombotic-vascular pathway and an inflammatory-neurotoxic pathway and argued that treatment should be individualized based on which pathway was true for a specific patient.[61] Citing evidence from mouse models, researchers concluded that cognitive and emotional impairment results from uncontrolled systemic inflammation, which throws off the balance between the immune and central nervous systems.[62] Moreover, a number of damaging factors then come into play, including elevated cytokine levels, aberrant immune cell activation in the brain, leakage of the BBB, and production of circulating antibodies cross-reacting with brain antigens.[62] Similarly, two different pathophysiologic processes contribute to neuropsychological events in SLE: an inflammatory pathway with an autoimmune-mediated cause and a thrombotic/ischemic vascular injury/occlusion process.[63] Often, a patient will have both processes at work simultaneously.

Lack of Understanding in Research

It is difficult to determine what causes cognitive impairment in lupus when we are still discovering what causes lupus. Researchers are discovering more details about how SLE negatively affects the body, such that there is a "loss of tolerance to nuclear antigens, the formation of autoantibodies, and immune complexes, resulting in complement activation, cell destruction, and tissue inflammation. (There are then) alterations in B-cell and T-cell activation, aberrant clearance of apoptotic material, and an activated type I interferon system."[63]

Recently, an article on a new understanding of disease pathogenesis concludes that it may "provide fertile ground for therapeutic development."[2] In that article, they discuss many possible pathways of etiology for SLE, including the following: "aberrant clearance of nucleic-acid-containing debris and immune complexes; excessive innate immune activation leading to overactive type I IFN signaling; abnormal B and T cell activation; multiple immune cells contribute to inflammatory amplification circuits; neutrophils activated by immune complexes; identification of new B subsets; disordered T cell regulation."[63] T cells of various types are, in fact, now being discussed as factors in many neurodegenerative diseases.[64] As mentioned previously, T and B cells are well accepted as part of the etiology of lupus. So, T and B cells should both be examined as a possible cause of neuroinflammation and cognitive impairment in autoimmune disease.

A recent article summarizes the pathogenesis of SLE by stating that "multiple cellular components of the innate and adaptive immune systems" are affected by the presence of autoantibodies and immunocomplexes, engagement of the complement system, dysregulation of cytokines and disruption of the clearance of nucleic acids after cell death.[65] Furthermore, an article in "Nature Immunology" cites B cell dysfunction as occurring early in lupus, with epigenetics (inherited phenotype changes) playing a significant part.[66] Studies note elevated B cell stimulator BlyS levels in patients with SLE, as well as many other circulating cytokines and complement activation.[67]

Balancing vs. Destroying

Dr. Joan Merrill, in discussing highlights from the 13th International Congress on Lupus, discussed new treatments that are now in development within trials and stated that we need treatments for lupus that **balance** the immune system, while current treatment tends to kill cells and destroy parts of the patient's immune system.[68] I completely agree with her philosophy, as it makes no sense to me to destroy a person's immune system through treatment, making a patient more susceptible to infections (I experienced immune-damaging side effects with Humira, for example.) Although I am not an immunologist or biochemist, the above articles give me hope that researchers are finally learning what is different at the cellular level in lupus. Hopefully, they will be able to utilize such cellular knowledge to develop more effective, specialized immune-balancing treatments in the near future.

Thankfully, this appears to be happening now, as researchers state that "recent insights into immunological mechanisms of disease progression have boosted a revival in SLE drug development."[69]

Waste Theory

As mentioned above, I find the "waste disposal" theory of lupus very interesting. The idea of waste involves apoptosis (cell death) and the inability of lupus patients to remove this type of normal waste from the body. As it builds up, it triggers autoantibody production and inflammation in the body, aggravating symptoms. There is speculation that genetic abnormalities predispose some individuals with lupus to these waste-clearing difficulties.[70] My recent DNA analysis revealed my body's inability to effectively remove toxins, decrease inflammation, and fight infections. My genetic SNPs have directly influenced these unfortunate characteristics (single-nucleotide polymorphisms–the most common type of genetic variation).

Autoantibodies

For whatever reason, lupus patients form autoantibodies that attack their healthy tissue. Sometimes, these autoantibodies fight against double-stranded DNA or ribonucleoproteins, etc.[71] The autoantibodies then form immune complexes that deposit in various organs, causing terrible symptoms (arthritis, rashes, nephritis, etc.). A study proposes a similar mechanism for cognitive dysfunction in lupus: "neurotoxic effects of autoantibodies directed at neuronal antigens such as N-methyl D-aspartate receptors (NMDAR), which affect neuronal function."[72] A group of researchers set out to link neuronal autoantibodies to NP+SLE, but neuronal-surface reactive antibodies were not detected, so they proposed future researchers look at innate immune complexes in detail for clues on the pathology of brain disease in lupus.[73] However, autoantibodies may not be the cause of cognitive dysfunction, as deficits may be due to "combined contributions of more than a single mechanism," such as cytokines, microglia, or neurotransmitters.[74]

Immunosenescence

As mentioned previously, a recent addition to the many theories of the etiology of cognitive dysfunction in lupus relates to accelerated immunosenescence or accelerated aging. Immunosenescence usually refers to normal aging processes that progress with age in our immune systems (both innate and adaptive). An increase in pro-inflammatory markers, like cytokines, is a large part of immunosenescence. In addition, chronic immune activation and inflammation change the immune system. All of this together, in lupus, causes accelerated immunosenescence.[75] These researchers believe accelerated immunosenescence correlates with cognitive impairment in SLE, particularly in attention, recall, and visuospatial tasks.[75]

Multifactorial Etiology

After reviewing evidence of neurophysiological changes, a group of renowned researchers made a broad conclusion, stating, "cerebral pathology is the critical factor in the development of cognitive impairment,"[22] Whereas another researcher more specifically concludes that "vascular abnormality, autoantibody and inflammation play important roles."[76] More specifically, it has been stated that "a role for autoantibodies, and molecular and cellular mechanisms in cognitive dysfunction" has emerged, "challenging our previous concept of the brain as an immune privileged site."[77] This was stated because the brain used to be thought of as a sterile organ, but that is no longer the case.

After reviewing the literature to date on neuropsychological symptoms of lupus, a study stated, "several pathogenic pathways are identified, such as antibody-mediated neurotoxicity, vasculopathy due to anti-phospholipid antibodies and other mechanisms, and cytokine-induced neurotoxicity."[54] Similarly, a group of researchers concede that the pathogenesis of NP+SLE is not fully known, but three factors are major contributors: pro-inflammatory cytokines, loss of brain barrier integrity, and autoreactive antibodies.[78] Recent research cites the "involvement of multiple aspects of the immune system in NP+SLE, including neurotoxic antibodies, pro-inflammatory cytokines and cell-mediated effects, in conjunction with abnormalities in neuroimmune interfaces," such as the BBB and choroid plexus.[79] Obviously, the central nervous system as a whole is directly involved in the pathogenesis of neurological symptoms.[1]

Another interesting study points out that deficits in various cognitive domains may even result from different etiologies or pathways.[56] Cuffari seemed to agree with the notion of various pathways, as she reviewed the literature on pathophysiology in NP+SLE and realized the presence of autoantibodies (ANAs, anti-NMDA, APS, etc.) may be a factor in NP+SLE development but also pointed out that increased cytokine production (like IL-1, IL-6, IFNalpha, TNFalpha) disrupting the BBB is likely another factor.[80]

Indeed, from the list of studies above, scientists are getting closer to understanding the myriad of factors involved in cognitive dysfunction in lupus. What is becoming increasingly clear is that it is, most likely, not one etiology but multifactorial.[81] This is quite similar to the complex pathophysiology of the disease itself. Indeed, scientists stated, "Our results suggest that many factors influence cognitive function in SLE" after finding in their study that disease activity and inflammation are two of those factors.[81] Researchers summarized the multiple factors driving cognitive impairment in lupus, including atherosclerotic changes, antiphospholipid antibodies, and chronic inflammation.[8]

Why do researchers put so much effort into determining the pathogenesis of cognitive impairment in lupus? A recent study provides an excellent answer: "Efforts are continuing in dissecting the pathogenesis of NP+SLE, as a better understanding of disease-relevant pathways will help in the development of more targeted therapies."[1] This is what we need—specific, targeted therapies that address the problem, but don't create new problems with immune suppression and side effects.

✶ ✶ ✶

In summary, cognitive impairment in lupus is a multifaceted issue arising from mechanisms such as systemic inflammation, autoimmune responses, vascular damage, neurotransmitter imbalances, and disruptions in neurogenesis. Despite ongoing research and advancements, the precise etiology remains unidentified, and it is evident that no single pathway can fully explain the cognitive deficits observed in lupus patients.

As understanding of these mechanisms deepens, there is potential to develop targeted therapies addressing these diverse pathways, thereby improving the quality of life for individuals with lupus. The interaction between the nervous and immune systems is complex, but ongoing research continues to advance towards finding effective treatments that can mitigate cognitive decline and provide clear recovery paths for those affected by this autoimmune disease.

Chapter 8

How Do You Identify Brain Fog in Lupus?

Lupus has come a long way from being seen as a rare and hopeless diagnosis, with new treatments offering more control over its physical symptoms. Yet, one critical aspect often goes unnoticed: its impact on the brain. Many people living with lupus struggle with memory lapses, trouble concentrating, and mental fatigue, which can deeply affect their daily lives. These cognitive issues are common and life-altering, making working, maintaining relationships, or even completing simple tasks difficult.

Despite growing awareness, there are still no clear answers or consistent methods for identifying, tracking, or treating cognitive dysfunction in lupus. This chapter looks at the ongoing need for reliable screening tools, the possibility of cognitive subtypes, and the search for effective treatments and biomarkers that could bring clarity and hope to those facing these invisible symptoms.

Brain Fog in Lupus: A Hidden Struggle

From the perspective of advancement, a study states, "lupus was once considered a rare disease with a universally fatal outcome...but the future is promising in terms of potential new treatments."[1] The remaining challenges stated in that article include improving the quality of life for patients by minimizing the use of corticosteroids, reducing infections and fatigue, and minimizing cardiovascular risks that still claim considerable loss of life.[1] I would add to these a resolution, or at least a proven, effective treatment for cognitive impairment, such that lupus patients can at least trust their brains, if not their bodies.

Indeed, SLE is "one of the most disabling autoimmune pathologies known to have an effect on the central nervous system secondary to the systemic disease."[2] Further, cognitive dysfunction in lupus is associated "with diminished quality of life."[3] It has also been stated that "up to 90% of people who have SLE have been found to have some problems with memory on formal (neuropsychiatric) testing."[4] Therefore, cognitive impairment is clearly not a minor issue.

Lack of Screening and Subtypes

Currently, most clinicians rely on lupus patients to self-report their cognitive deficits, which usually doesn't happen until the impairment is substantial, if at all.[5] "There remains a need for an objective and non-invasive means to establish neurological dysfunction in SLE patients, both for investigative purposes, as well as for clinical diagnosis and surveillance."[6]

To address this gap, we first need efficient, cost-effective screenings and tests to identify specific deficits and track changes over time. These screeners or tests must be consistent across studies and clinics to measure the same thing within and across patients. Along with screeners/tests, we need to find the best brain imaging techniques to monitor neurophysiological changes.

Once that is all in place, research should focus on determining whether a specific profile (or multiple profiles) of cognitive deficits affects patients with SLE or whether types of deficits are truly individualized in each patient. It has been postulated that subtypes of cognitive impairment may exist within this population.[7] These subtypes may be investigated through case studies looking at the interaction between genetics and environmental factors.[8]

Moreover, researchers should also study the course of cognitive impairment over time in patients (whether there is improvement, regression, or stability) and what the confounding factors may be (such as depression/anxiety, other diagnoses, medications, etc.). If researchers could find biomarkers tied to impending flares, physicians could perhaps prevent the disease from worsening with targeted treatments.[9]

Standardized Biomarkers Needed
Researchers advocate for further research on biomarkers specifically tied to cognitive dysfunction in SLE, as they state potential factors may include anti-NR2A/B antibodies, MMP-9, NETS, and other pro-inflammatory factors.[10] Similarly, another group of researchers advocate for further biomarker research and provide direction for biomarker research, stating we need a clinical standard for biomarkers and imaging in lupus.[11] For example, research with functional MRI found functional connectivity between the basal ganglia and the cerebellum (vermis 6) "may serve as a neuroimaging marker for evaluating the progressive cognitive decline in SLE patients."[12]

In addition to these insights, we also need future research connecting certain biomarkers with different types of treatment.[13] Perhaps most importantly, research also needs to focus on specific evidence-based treatment for cognitive impairment to reverse the symptoms and improve daily performance. Hopefully, an increased understanding of pathophysiology, from research like that described above, will lead to information on how to prevent cognitive deficits altogether. The ultimate hope is more treatment options for lupus, in general, and, eventually, a cure for this horrible disease.

Screening, Testing, & Imaging
Cognitive dysfunction is a significant but often underrecognized complication of systemic lupus erythematosus (SLE). Despite longstanding awareness of its impact on a patient's quality of life, the medical community struggles to identify, define, and systematically address these cognitive challenges within standard clinical practice. Although comprehensive neuropsychological assessments have been proposed and even formalized by

professional organizations, their complexity, cost, and limited accessibility make them impractical for widespread use.

As a result, there is an ongoing search for reliable, efficient, and affordable screening tools that can be integrated into routine rheumatology care. This chapter explores the current landscape of cognitive assessment in lupus, highlighting the limitations of traditional methods and examining emerging alternatives that may offer more practical solutions for early detection and intervention.

Challenges in Cognitive Assessment

As stated by some researchers, "an important component of medical care for SLE patients is a thorough assessment for neurocognitive complications."[5] However, I don't see this being done in typical rheumatology clinics. We also have no consistent, agreed-upon measure of cognitive dysfunction of lupus.[7] After a systematic review and meta-analysis, it has been concluded that we need better "identification and definition of cognitive deficits" in lupus patients and "adequate cognitive remediation programs."[14]

In addition to this, you may recall from earlier in this book that the ACR recommended a "gold standard" test battery to diagnose cognitive impairment in SLE patients.[15] The ACR test battery includes all the following: "Revised Wechsler Adult Intelligence Scale; Digit Symbol Substitution Test: The Trail Making Test Part B; The Stroop Color and Word Test; The Learning Trial from the California Verbal Learning Test; The Immediate and Delayed-Recall from the Rey-Osterrieth Complex Figure Test; The Controlled Oral Word Association Test; Animal Naming Test; The Finger Tapping Test."[16] Moreover, in 2007, a Cognition Sub-committee of the Committee on Lupus Response Criteria recommended the ACR-NB, consisting of 10 neuropsychological tests.[17]

As you can imagine, test batteries such as these are costly, since you must pay an independent neuropsychologist or a speech-language pathologist to administer them. The test battery is often not covered by insurance. Also, some patients find it to be inconvenient and exhausting, as it generally takes multiple hours, and it may even be completely unavailable, particularly in rural areas. Since the battery of tests recommended by the ACR is costly, time-consuming, and requires specialized training, the need for other tests and screeners is apparent.[18] In conclusion, "neuropsychological tests are very time-consuming in standard clinical practice, suggesting an urgent need to design a shorter cognitive screening test that can easily be used by busy practicing clinicians."[5]

Standard Screening
Many other screeners and tests have been proposed as more feasible options to identify cognitive impairment in lupus. In fact, "there have been several alternatives to the neuropsychological battery that have been validated and have shown clinical utility in the rheumatic population (particularly SLE)."[17] The most frequent tool administered as a cognitive screening is the Mini-Mental State Examination (MMSE). However, this tool is insufficient for MCI detection, as discussed below.

While developed as a screening for Alzheimer's, the MMSE has been utilized occasionally in lupus research. One study found a positive correlation between MMSE scores in patients with SLE and fMRI connectivity values, showing a decline in communication between the basal ganglia and cerebellum.[12]

Another type of comprehensive test is the CNS Vital Signs Test, which is a test battery consisting of seven established tests which is given by computer.[19] This test can be taken from home or in a clinical setting.

For another option, one study used the Cambridge Neuropsychological Test Automated Battery (CANTAB) to assess cognition in lupus patients, which is an interesting option since it is both nonverbal and given by computer.[20] The CANTAB would be a good international option because of the nonverbal component. Still, these comprehensive evaluations are not usually part of standard treatment at most rheumatology clinics and are often not covered by insurance benefits. Therefore, there remains a need for an affordable, readily available screener.

Testing Battles

There is an unmet need to find the best screening tool for cognitive impairment in lupus patients.[21] Further studies are needed to determine which screeners or tests are best for identifying cognitive deficits in lupus but, at the same time, have a lower cost.[22] Likewise, a similar search for the appropriate screener has been underway for multiple sclerosis (MS) patients.[23] A few recent studies have been conducted to determine which screeners might be the best choice for lupus specifically. One study suggested the Cognitive Symptom Inventory (CSI) or the automated computer-based neuropsychological assessment metrics system (ANAM) as good options.[24] Another study assessed 211 lupus patients utilizing both the ACR test battery and the ANAM screening and found good validity for ANAM identifying cognitive impairment.[25] They even found that they could shorten the ANAM down to 20 minutes (using certain subtests) and still effectively identify those patients with cognitive difficulties.[25] Additional research is needed to verify if ANAM identifies the same patients with cognitive issues as the ACR-SLE suggested battery while having an "acceptable administrative burden."[21]

Furthermore, a recent study found SLE patients to score significantly worse across all domains of the ANAM when compared to healthy control subjects.[26] Researchers compared three screening tests against the ACR neuropsychological battery (mentioned above) and found the Montreal Cognitive Assessment Test (MoCA) to be the best option.[27] The MoCA is definitely more challenging than the MMSE.[28] Remember, MMSE is the most utilized screener in doctors' offices. The MoCA showed the highest correspondence with the original test battery for lupus, and researchers have concluded it is highly effective for detecting cognitive impairment.[27]

Although they assessed patients over 60 from the general population rather than autoimmune patients, a recent systematic review found the MoCA to be better at the detection of MCI than the widely utilized MMSE (MiniMental State Examination). A systematic review of 39 studies looking at 42 tests for MCI found the MoCA was the best test with high sensitivity for aMCI (amnesic mild cognitive impairment- focused on memory deficits) and also had good test-retest reliability.[29] Specifically, one study found lupus patients to have significantly lower scores on the MoCA and the MMSE than healthy control subjects.[30] Another study, which selected lupus patients without known CNS symptoms (NP-SLE patients), found **all** the patients had cognitive impairment as assessed by the MoCA.[31] The MoCA takes only about ten minutes to administer, so it could be readily available to patients at rheumatology clinics as long as a qualified clinician was available to administer the test.

A different study compared two screening tests for cognitive impairment in SLE: the MoCA and the IQCODE (Informant Questionnaire on Cognitive Decline in the Elderly-a questionnaire filled out by a family member) and found, like the study above, the MoCA is a "promising and practical screening tool."[32] They also compared these two tools to the ANAM (Automated Neuropsychological Assessment Metrics), completed on the computer, and found six out of ten MoCA questions to be correlated with the ANAM total score.[32] Therefore, MoCA may be an effective and practical screening option for lupus patients.[33]

In addition, another study utilizing MoCA found the prevalence of MCI to be 48% in lupus patients.[34] These researchers go on to discuss how the lack of appropriate and available cognitive screening tools for lupus patients probably delays the diagnosis and monitoring of cognitive impairment in this population.[34] A group of renowned researchers state mild cognitive impairment precedes dementia in SLE patients, and the MoCA is an "ideal tool" for identifying MCI in the population.[35] Further, remote assessment of the MoCA is now available, and remote neurobehavioral assessment has become more accepted recently due to the pandemic.[36] The combination of the MoCA with a patient-reported tool, such as the Cognitive Symptoms Inventory, may also be considered as an even more complete option.[37]

Additional Options

Another screening option was recently discovered by researchers in a spatial navigation task (SNT).[38] The SNT task was so sensitive it not only identified lupus patients with cognitive dysfunction but poor performance on this task was also correlated with hypometabolism in certain brain regions (as measured by PET scans), as well as blood titers (high serum DNRAb).[38] Moreover, research has also found another test, the Paced Auditory Serial Addition Test (PASAT), to reveal working memory deficits in SLE patients, and they concluded dysfunction of frontal white matter might precede abnormalities that could be seen on MRI.[39]

A new option recently arrived on the scene, as researchers from Kanazawa University have created a new computerized assessment battery for cognition (C-ABC), which they report can distinguish individuals with MCI from people with normal cognition in only five minutes.[40] If verified by independent studies, the ease and efficiency of this tool, not requiring significant expense or trained administrators, would erase many of the obstacles to screening autoimmune patients for cognitive impairment within a rheumatology clinic.

Furthermore, in an eight-year study following 665 people (not lupus patients), the SAGE (Self-Administered Gerocognitive Examination) screener was able to identify patients with mild cognitive impairment accurately and has the advantages of having four interchangeable forms and taking only 10-15 minutes to self-administer and one minute to score. By not requiring a clinician or examiner to administer, an option like this would save time and money in a clinical setting.[41] However, it can only be possible if it is found to be valid for lupus and other autoimmune patients. There is a digital online version of the SAGE called BrainTest. When selecting an online screener for an individual, however, the computer skills of the person must be considered.

The National Institute of Health (NIH) has a free, computerized test battery to assess cognition, called the NIH Toolbox Cognition Battery, which could be explored as a tool for lupus patients in research use.[17] More in-depth research on cognitive testing with lupus patients, specifically, is needed.[42] This should to be done with some universal agreement on appropriate screening techniques or tools. Further, the American Academy of Neurology (AAN) recommends that providers (regarding the general population) "assess for MCI with validated tools in appropriate scenarios" and that clinicians should monitor cognitive impairment over time.[43]

From the studies discussed above, whether the MoCA, the C-ABC, the SNT, or even an undiscovered tool may be the best method for screening and monitoring cognitive deficits in lupus patients is unclear. For a more in-depth review of four objective tools (ACR-NB; ANAM; HVLT-R; COWAT), which can be utilized to assess the cognition of lupus patients, please see Yuen, et al. (2020). The four measures of cognition they reviewed were found to "have psychometric evidence for reliability, validity, and responsiveness generally and in rheumatic disease (mainly in SLE)."[17] As it has been observed that "the establishment of an effective diagnostic protocol is of clinical importance, since NP+SLE is not a rare problem."[44]

Subjective Measures

Some researchers have found subjective measures (like patient questionnaires) do not always correlate with objective testing.[21] Similarly, another study found that self-reported cognitive symptom measures were less sensitive to cognitive deficits than standardized tests.[45] However, a patient's perceived deficits must be assessed and treated as valuable in their own right, as all patients deserve deep listening and respect from their providers.

In addition, I suspect patients would be more open to a questionnaire-based screening as opposed to being assessed by an examiner, since there may be less embarrassment. In one study, 277 SLE patients were given questionnaires, and 41.7% self-reported problems with forgetfulness, while 29.5% reported difficulty concentrating.[46]

The U.S. Food and Drug Administration has emphasized the importance of patient-reported outcomes (PROs). One good way to assess these outcomes is through reliable and valid questionnaires/surveys, such as the Lupus Quality of Life Questionnaire (LupusQoL) and the Functional Assessment of Chronic Illness Therapy-Fatigue Scale (FACIT-F).[47]

Some researchers claim they have been unable to find a questionnaire that correlates well with the gold-standard ACR battery. Patient outcomes and quality of life may not, in fact, directly correlate with standard scores on objective cognitive tests. Therefore, both types of measures (subjective and objective) are likely necessary for a complete picture. In fact, future clinical practice should include both types of evaluation of cognition at the initial diagnosis of SLE and throughout the course of the disease to track changes at all levels (Rayes et al., 2018).[21]

To summarize, although more research is needed, we have plenty of options available to begin routinely assessing cognition in patients with lupus (and other autoimmune diseases) at rheumatology clinics everywhere.

Confounding Factors

As an aside, we must mention confounding factors in diagnosing cognitive impairment in SLE. Cognitive deficits from CVAs/strokes, traumatic brain injuries, aneurysms, Alzheimer's, etc., must be separated from general cognitive impairment in lupus. Patients with these types of diagnoses are typically excluded from research studies, attempting to isolate patients with SLE and associated cognitive impairment only. Other factors, like infections, metabolic complications, and drug toxicity, can cause neurological or neuropsychiatric symptoms that must also be weeded out from SLE causes.[48]

In this regard, scientists utilize various attribution models to make sure the symptom is attributable to SLE, but some of these models omit "minor NP (neuropsychiatric)" events, so they may actually leave out mild cognitive impairments.[49] Some attribution models are very complex, and one group of researchers even developed an app called "neuro lupus" to help physicians determine whether neurological symptoms in SLE patients are, in fact, attributable to lupus.[50] While some researchers have searched for a similar pathway or etiology for lupus cognitive impairments and general dementia or Alzheimer's, many are convinced the etiology is entirely different and that autoimmune disease uniquely causes cognitive deficits. In fact, one study suggests the cognitive changes in lupus "represent an SLE-specific pathology irrespective of disease activity and other confounders such as medications or prior central nervous system events."[51]

In addition to confounding factors related to previous neurological events or additional medical diagnoses, we must consider the influence of other primary symptoms. Lupus patients do not simply have cognitive impairment, but they may also have pain, inflammation, fatigue, mouth/nose sores, rashes, fevers, etc. To what degree, if any, do those other symptoms influence cognition? A group of researchers stated, "CI (cognitive impairment) tends to develop insidiously over the course of the disease, independent from other SLE manifestations."[51]

In contrast, other researchers think symptoms influence and overlap. "Inflammation and depression may suppress brain response, and as these improve, brain responses start to "normalize."[52] Interestingly, studies talk about how the right insula of the brain is significantly affiliated with pain, and they found decreased nodal efficiency (capacity for communication) in this region in lupus patients.[12] Since lupus patients have chronic pain and a decreased amount of N-acetyl aspartate (NAA) in that location (right insula), which has been associated with neuronal damage and loss, could future research show a connection between constant pain and cognitive difficulty? Intuitively, this makes sense, since none of us excel at cognitively-challenging tasks when in serious pain. For example, have you ever tried to grade college essays with a migraine? Managing various symptoms requires patients to balance daily tasks, and choose the appropriate time for each.

Emotional factors and stress levels should also be considered on an individual basis. A clinician may consider adding psychotherapy as a supporting aid to cognitive treatment. There are high levels of depression and anxiety in autoimmune populations, as these patients have to deal with multiple symptoms and unwanted life changes. A review of patients with autoimmune thyroiditis, one of the most common autoimmune diseases, for example, revealed significantly higher anxiety and depression scores.[53] A psychologist or counselor could bolster the patient's ability to deal with unexpected life changes by helping them process change and teaching them positive stress-management techniques. In addition, a certified health coach can help a patient manage challenges in a healthy way and make positive changes to their lifestyle. A team approach is often best for autoimmune patients.

Finally, we must acknowledge the confounding factors in the treatment. Cost has already been mentioned as an impediment to both testing and treatment. In addition to cost, however, a patient's willingness to change can affect treatment. Lifestyle changes, whether dietary, exercise, or brain training, require a person's willingness to change their daily routine. Friends and family members should encourage healthy behaviors, yet support the patient's wishes regarding such changes. Professionals must be ready to offer services and information on best practices, while respecting the patient's final decision regarding their choices.

Levels of Impairment

Another issue with research on cognitive impairment thus far is that there has been no agreement among lupus researchers as to what defines mild cognitive impairment/dysfunction versus more severe cognitive impairment or dementia. Once there is agreement on specific cognitive screeners or tests to be utilized consistently across clinics and across time, standards must be set on how to classify lupus patients as having cognitive impairment at various levels. This categorization may be done with labels or levels or simply scores on a scale.

In this book, I have been unable to separate the levels of cognitive impairment in the research, so I have chosen to include research on cognitive impairment, cognitive dysfunction, and MCI, but I believe it is the milder cases that often go ignored and untreated (and even untested) in the real lives of people with lupus.

Indeed, a group of researchers said that "small changes like mild cognitive impairment are often left unnoticed."[54] Studies have used various types of tests and screeners to assess cognition in lupus patients and defined "cognitively impaired" or "cognitive dysfunction" in various ways. For example, one study found significant cognitive impairment tied to brain atrophy in two regions and classified such impairment as "mild" if the patient had deficits in fewer than three cognitive domains.[55] I think it's absurd to think having deficits in two types of cognition is mild, when you consider how this affects the person's life.

I would hypothesize that some of the patients not classified as MCI but struggling in one or two cognitive areas (executive functioning and language, for example) may have been classified as having cognitive impairment in other studies (according to other criteria) and most likely **do** have significant cognitive deficits that are negatively impacting their daily lives.

Screening of Disease Activity
In addition to screenings to identify and track cognition, we need excellent screening options to measure overall disease activity over time in lupus patients. For an excellent review of the most common screeners for disease activity, see Romero-Diaz, et al. (2011).[56] They conclude reliable measurements of disease activity can help set baseline levels of activity, determine treatment effectiveness,

encourage shared decision-making for treatment, and help develop protocols for treatment, but they point out that the multi-system nature of lupus and the presence of flares makes things more difficult.[56]

Moreover, the best screeners should also track cognitive functioning over time amid more obvious physical symptoms. Complicating the issue, a group of researchers discovered that cognitive and physical functioning levels did not always change along with disease activity and/or disease damage, which suggests that disease activity screeners will not capture all levels of patient functioning and, thus, might miss cognitive changes.[57]

In contrast, a study found relationships between cognition (measured by the Self-Administered Gerocognitive Exam or SAGE) and higher lupus-related damage (measured by the SLICC-DI), so some general lupus screeners might indeed tie in with cognition.[24] However, the SLICC-DI was not predictive of neuronal damage since the mechanisms of tissue damage in the CNS differ from tissue damage in peripheral organs.[58] Therefore, more research is needed to clarify this association.

Brain Imaging Options
Further research is needed to determine which imaging techniques might best identify areas of damage in the brains of patients with lupus, which correspond with various cognitive difficulties. Imaging has the potential to be an effective noninvasive biomarker for cognitive impairment in lupus since "no effective non-invasive biomarkers are available to evaluate the cognitive status of patients" as of yet.[12] Unfortunately, brain imaging is often not part of a routine workup of an SLE patient, so an underestimation of CNS difficulties may exist in the lupus population.[59]

In the previous chapter of this book discussing the neurophysiology of cognitive dysfunction in SLE, you may have noticed various imaging tools were utilized. In fact, "the recent advent of structural and functional neuroimaging of the brain has been recognized to be promising in detecting, monitoring and possibly predicting lupus-related CNS damage."[60] Moreover, a study performed a systematic review of various imaging studies on lupus patients.[61] MRI (magnetic resonance imaging) remains the "neuroimaging technique of choice" for neuropsychiatric events in SLE and is the imaging technique most often used in clinical practice.[62]

Moreover, a study recently proposed a "3D arterial spin labeling" technique with MRI for early detection of neurological lesions in lupus patients.[63] One group of researchers concluded, "MRI is the best imaging technique for the non-invasive diagnosis of NP+SLE, even though a wide range of nonspecific abnormalities have been reported."[30] Although less available in actual practice, some SPECT studies show hypoperfusion (less regional blood flow) in the frontal lobes of lupus patients with cognitive deficits.[64] Still, other studies have found no association between cognitive dysfunction and decreased cerebral blood flow.[65] In addition, computerized tomography (CT) scans are not often used in this body of research since they cannot detect structural brain abnormalities as well as MRI scans do.[55] One study found f(functional)MRI was able to indicate working memory deficits and may be a good choice as a more reliable indicator of neuronal damage in lupus patients.[58]

MRI and fMRI studies have been utilized frequently in the last decade of research to shed some light on what physiological brain changes correspond with cognitive changes in lupus. One study, using MRI, found greater white matter (deep, axonal tissue) lesions in all SLE patients (NP-SLE and NP+SLE) compared to healthy controls and discovered the volume of lesions correlated to cognitive dysfunction.[66] Another MRI study found a higher frequency of white matter changes in NP+SLE patients, but changes were also found at a lower rate in NP-SLE patients.[67] More specifically, diffusion tensor imaging (DTI) found definite white matter integrity changes in the prefrontal cortex of both NP+SLE and NP-SLE lupus patients.[59]

Another study utilizing MRI found both gray matter volume and white matter volume reductions in SLE patients, with an obvious association between cognitive impairment and white matter volume reduction.[68] Similarly, it has been hypothesized that SLE patients with greater cognitive deficits may have both gray and white matter involvement in their brains.[69] Moreover, research has also concluded that standard MRI was not sufficient to either establish or exclude NP+SLE diagnoses.[70]

Some researchers think we need more sensitive MRI techniques to find neurophysiological changes in patients with cognitive deficits.[71] This is because basic MRI scans cannot always detect subtle white matter hyperintensities. Researchers recommend a technique called diffusion-tensor MRI, which can actually map the brain's white matter tracts and may lead to biomarkers of microstructural integrity of the brain.[72] In fact, recently, it has been found that classic MRI machines identified lesions in the brains of only 50% of NP+SLE patients, while more advanced MRI techniques can detect even pre-clinical lesions.[73]

Most researchers agree we need more advanced MRI techniques, in addition to conventional MRI, in order to detect subtle changes in the brain.[74] This is the case because brain changes may be obscure. One study suggested that event-related potentials could also be used as a complementary tool in assessing cognitive impairment in lupus patients.[75] A 2012 study discovered that SLE patients had increased cerebral blood flow and cerebral blood volume compared to healthy subjects, with the most notable differences in the posterior cingulate gyrus for those with active disease.[76]

After reviewing the literature on PET (positron emission tomography) and SPECT (single-photon emission computed tomography) imaging with lupus patients, researchers stated: "it appears that hypo- and hypermetabolism may relate to two different yet coexisting aspects of brain involvement in SLE."[62] Moreover, it has also been concluded that PET and SPECT have higher potential, as compared to MRI, for picking up neurological abnormalities in lupus.[77] For example, the micro-thrombi that have been found and linked to NP+SLE may be visible with PET but not with MRIs.[78] Other researchers think fMRI is better than MRI or SPECT because it is highly sensitive to finding brain areas tied to more subtle cognition dysfunction.[79] Alternatively, a better option may be to combine MRI and SPECT results.[80]

Brain damage is thought to progress over time as cognitive dysfunction increases, from white matter hyperintensities to brain volume loss or atrophy.[81] This may be why we don't see uniform results across scans. In fact, no differentiation in white matter lesions was found between NP-SLE and NP+SLE subgroups, as stated above, but the NP+SLE subgroup did differ in atrophy of the bilateral hippocampus, which correlated to greater cognitive deficits in the NP+ group.[82] The researchers concluded that "neuroimaging and its recent applications are of utmost importance for diagnosis and monitoring" (SLE).[83] Moreover, researchers also state cognitive dysfunction (as well as anxiety, mood disorders, etc.) are "diffuse symptoms" that are much harder to identify accurately with MRI, as opposed to "focal symptoms" such as stroke, seizures, etc.[84]

In addition, a specialized neuroimaging technique, MR spectroscopy, was found to correlate cognitive dysfunction in lupus with abnormalities of the choline-to-creatine ratio, which shows white matter integrity.[85] (For a more in-depth look at various techniques, see Zardi, et al. (2014), who reviewed the strengths and weaknesses of modern neuroimaging.)[86] A group of well-known researchers states that we need "multimodal imaging" with various imaging techniques to be the future potential standard of practice in neuropsychiatric SLE.[62] Moreover, it has been pointed out that our imaging techniques are not "sufficiently sensitive or specific enough at this time."[33] One thing is clear: to date, we have insufficient knowledge of the exact pathophysiology of cognitive impairment in SLE, and we definitely need the appropriate tools for this discovery.

As discussed above, appropriate screening and testing are needed to detect autoimmune patients' cognitive deficits. Perhaps future research needs to identify a specific combination of paper/pencil or computer cognitive tests, combined with a protocol of neurophysiological testing and/or imaging, in order to obtain a complete picture of a patient's deficits. This combination protocol could then be repeated periodically to track cognition over time and, hopefully, to guide specific treatment for deficits. Indeed, researchers suggest that "a combination of emerging screening tools, biomarkers and imaging techniques is being utilized to improve how NP+SLE patients are evaluated."[86] While a combination of tools may be the norm in research, I question whether we are actually seeing these techniques utilized in the reality of clinical practice.

In conclusion, cognitive impairment in lupus remains a significant yet under-recognized challenge that demands greater attention from both researchers and clinicians. Although treatments for the physical symptoms of lupus have advanced, cognitive dysfunction continues to undermine patients' quality of life. The development and adoption of reliable, standardized screening tools are essential to identify and monitor cognitive deficits early and accurately. Future research must focus on clarifying the presence of cognitive subtypes, understanding the progression of cognitive changes, and identifying biomarkers that could predict or prevent worsening symptoms. Both objective and subjective measures are necessary to create a comprehensive understanding of patients' experiences and outcomes.

Additionally, addressing confounding factors such as pain, emotional health, and other overlapping symptoms will be critical in improving cognitive care. A multidisciplinary approach that includes medical professionals, mental health providers, health coaches, and patient support systems can help manage the complex needs of individuals living with lupus. Ultimately, consistent cognitive screening and proactive management have the potential to significantly enhance the daily lives and long-term outcomes of lupus patients. By prioritizing cognitive health alongside physical treatment, the future of lupus care can become more holistic, compassionate, and effective.

Chapter 9

What About Domains, Demographics, & Quality of Life?

Cognitive dysfunction is a significant and often overlooked aspect of systemic lupus erythematosus (SLE), affecting patients' daily lives in profound ways. Despite advances in clinical research, there is still considerable debate regarding which cognitive domains are most impacted by lupus. Studies have shown varying patterns of cognitive deficits, with areas commonly affecting attention, memory, language, and executive function. The role of demographic factors, such as education and gender, in influencing these cognitive changes is also under investigation, with some studies suggesting that males and those with less education may be more prone to cognitive impairments.

Furthermore, the impact of cognitive dysfunction extends beyond health, influencing quality of life and employment outcomes. This chapter delves into these issues, exploring how cognitive dysfunction in lupus affects patients' functioning and well-being and highlights the urgent need for further research and effective treatment strategies.

Cognitive Domains
A recent study stated, "the convergence of more rigorous clinical characterization, validation of biomarkers, and brain neuroimaging provides opportunities to determine the efficacy of novel targeted therapies in the treatment of NP+ SLE."[1] At the same time that the

problem of affordable and efficient screening/testing/imaging options to identify and follow cognitive problems is being investigated, the research literature is still in disagreement regarding which areas or cognitive domains are most affected by SLE. No one has identified a specific pattern or profile for cognitive deficits in lupus patients.[2] Therefore, there is no one accepted pattern of cognitive domains affected for all SLE patients.[3] A study, for example, found problems in language, attention, and delayed recall.[4] Another group of scientists found visuospatial and executive functions to be the most negatively affected cognitive domains.[2]

Following a meta-analysis, it was concluded that patients with SLE had statistically significant deficits in visual attention, visual memory, and visual reasoning.[5] Similarly, it had also concluded that patients with SLE (with or without NP+SLE) demonstrate executive functioning deficits (compared to controls), such as cognitive fluency and reasoning.[5] They further deduced all SLE patients had reduced visual attention and visual learning (immediate visual memory), and the researchers stated they were surprised to find reduced performance in the language domain.[5] (*Attention speech-language pathologists:* I believe that language is an under-studied area in the autoimmune population and requires further research.)

Moreover, the three subdomains SLE patients struggled most with in one study (as measured by the MoCA) were attention, delayed recall, and language.[4] Further, Lampner suggests there might even be phenotypes of MCI in lupus patients, which vary by pathogenesis (disease development).[6] A recent study dividing SLE patients into two groups based on symptom burden seems to verify this, as they found that subtype A performed worse on objective cognitive tests and also had greater levels of reported subjective cognitive impairment than group B.[7]

In contrast, no differences between cognitive domains affected were found when SLE patients' test results were divided into mild, moderate, and severe cognitive impairment profiles.[8] There may be one profile of deficits, then, or many profiles, or, in fact, no common profiles or lists of difficult cognitive areas, but simply individualized deficits on a patient-by-patient basis. Based on my clinical work with stroke patients, for example, there are common cognitive profiles depending on the area of focal lesion, but even then, each patient is unique within that profile for their specific cognitive deficits. Since lupus patients are more likely to have diffuse areas of affected brain regions, I would expect each patient to have a unique cognitive profile or various deficits specific to them. Researchers suspect better "sub-grouping is required for clinical and research purposes."[9] This is true for research purposes, as in dividing subjects for a study, but I believe grouping may not be as useful for clinical treatment in this population. Each autoimmune patient with cognitive deficits should be assessed, and then treatment should be tailored to the deficits revealed for that person.

Demographic Factors

Another area that warrants study is whether demographic factors play a role in cognitive deficits in lupus patients.[10] One study found that patients with less education had lower cognitive scores.[4] At the same time, other studies have not found education level to be significant.[11] Similarly, research found males with SLE in their study had a greater prevalence of cognitive impairments (62% versus 33% in females).[12]

In addition, risk factors (if applicable) for cognitive changes in lupus should be examined (such as flares or certain medications or psychological conditions). For example, some researchers have hypothesized that corticosteroids–commonly used in lupus patients–are the actual cause of cognitive deficits. However, cognitive symptoms of decreased psychomotor speed deficits, immediate recall, and complex attention/executive function have been found in SLE patients who had not had corticosteroids and had never been previously diagnosed with neurological symptoms.[13]

Quality of Life

We also need more research studies to investigate the impact of cognitive dysfunction on patients' quality of life (QOL) and vocational outcomes. QOL was improved in a recent study in which SLE patients had decreased pain when they were taught how to practice the Benson relaxation response technique (a method to relax muscles in your body while breathing and clearing your mind).[14] While this is useful, QOL goes beyond pain levels, and some autoimmune patients have to worry about whether their cognitive abilities will allow them to complete everyday work and household tasks.

In a study with 205 lupus patients, even those with low disease activity had both activity impairment and work productivity issues, both negatively affected reported QOL.[15] Moreover, three types of stigma can persist with cognitive impairment: self-stigma, affecting a patient's view of themselves; societal stigma (which affects employment); and clinician-related stigma, with a third of clinicians believing nothing can be done to help.[16] Lampner states, "cognitive dysfunction in patients with SLE is associated with substantial negative effects on functioning, employment potential, and quality of life."[6] Indeed, research has shown a definite association between cognitive ability and unemployment.[17]

Impaired cognition impacts each individual differently depending on which cognitive domains are affected and the severity levels. One recent disturbing study concluded that cognitive dysfunction was highly prevalent in lupus but had a low impact on quality of life.[18] This is absurd, and upon closer examination, they admitted that they eliminated patients from the study with severe presentations, and they also did not measure subjective or more subtle cognitive impairment.

Cognitive impairment definitely affects a patient's quality of life. It can affect their health and happiness by negatively influencing how they are able to work, parent, play, and enjoy life. Supporting this, Mackay firmly states cognitive impairment, fatigue, and depression have debilitating effects on the quality of life of lupus patients.[19] Finding deficits in executive function (particularly cognitive flexibility, attention, and processing), researchers pointed out that executive function is necessary for work, particularly in jobs that require high cognitive performance.[20] In fact, cognitive impairment can lead to unemployment and a poor quality of life.[21]

Memory impairment has been associated with employment status in lupus patients, such that 56% of patients with severe memory impairment are unable to work at all.[22] In addition, cognitive deficits can lead to depression, anxiety, and isolation as the patient becomes less confident in social situations. Harrison states common consequences of cognitive dysfunction in SLE patients include the following: "disability, distress, daily activities (ADLs), self-esteem issues, isolation, decreased quality of life, poorer medical compliance, and worsening of other lupus symptoms." [23]

The disease of lupus, in general, often leads to disability, and it is impossible to predict which patients will be unable to work at some point after diagnosis. Research is beginning to explore this factor, as one study found that an EULAR/ACR diagnostic score of greater than or equal to twenty predicts a higher disease activity five years post-diagnosis, more organ damage, and a greater mortality rate.[24] The bottom line is currently, "there are no specific recommendations to prevent its (cognitive dysfunction) onset and no defined treatment once it is diagnosed."[25] The lack of prevention and treatment remains in spite of the fact that cognitive dysfunction frequently occurs in many lupus patients, and it negatively affects social functioning, work capacity, employment, and quality of life. I can attest to this in my own life, and I feel that the current lack of **evidence-based** prevention and treatment options is inexcusable. That said, there are options for treatment and improvement, which will be covered in the following chapters.

✶ ✶ ✶

Cognitive dysfunction in patients with systemic lupus erythematosus (SLE) significantly affects their quality of life and employment outcomes, yet remains an under-researched and poorly understood aspect of the disease. The impact on cognitive domains such as attention, memory, language, and executive function is profound, but the variability across individuals makes it difficult to pinpoint consistent patterns. While some demographic factors, like education and gender, may influence cognitive impairments, the exact nature of this relationship remains inconclusive. The lack of a clear profile for cognitive deficits in SLE patients underscores the need for more tailored research, particularly in identifying subgroups with different cognitive challenges.

Moreover, the consequences of cognitive dysfunction extend beyond health, with marked effects on work productivity, daily functioning, and social interactions. The stigma associated with cognitive impairment further compounds these challenges, creating barriers to effective intervention and support. As research continues to explore potential causes and treatment options, it is clear that more focused efforts are required to address this aspect of SLE. While researchers continue to evaluate prevention and treatment strategies, it is essential to prioritize the early identification and management of cognitive dysfunction in each individual now, in order to improve their overall quality of life. This issue must not be overlooked any longer, as it has a lasting impact on the well-being and employment potential of those affected by SLE.

Chapter 10

What Are The Cognitive Treatment Options?

Cognitive dysfunction in systemic lupus erythematosus (SLE) remains a complex and often overlooked challenge, with no universally accepted treatment to date. While early studies declared a lack of effective interventions, more recent literature has begun exploring promising paths, including pharmacologic strategies and cognitive rehabilitation. Though pharmaceutical approaches such as cholinesterase inhibitors or agents that support the blood-brain barrier have been explored in broader populations, their application in lupus-related cognitive impairment is still uncertain. Meanwhile, cognitive rehabilitation—once doubted as viable for progressive deficits—has gained recognition for its potential benefits, especially in patients with mild impairments.

This chapter aims to explore current research, evolving perspectives, and the growing body of evidence supporting cognitive interventions. It also highlights the critical need for greater awareness among healthcare providers and patients alike and the role of speech-language pathologists in offering treatment. With limited access and variable support from insurance or local services, alternative methods like telepractice are also considered, pointing toward an emerging, though still developing, landscape of treatment options.

Cognitive Treatment in Lupus

Regarding the treatment, Harrison stated that "there is currently no treatment available or recommended for cognitive dysfunction in lupus patients."[1] On the other hand, research states two treatment approaches to cognitive dysfunction in lupus could be considered–cognitive rehabilitation and pharmacologic treatment–but neither had been established as efficacious at that time.[2] A group of researchers stated, "no treatment for SLE-mediated cognitive dysfunction exists," but they go on to speculate future pharmacological treatments may include those that inhibit microglial activation, like an ACE (angiotensin-converting enzyme) inhibitor or those that protect the BBB integrity.[3]

In addition, upon reviewing the literature on cognitive treatment with the general population of people with "amnestic MCI"–mild cognitive impairment with memory issues–a study found administering cholinesterase inhibitors to have some benefit, but it remains controversial.[4] Even more severe cognitive impairments, like Alzheimer's, have few effective pharmaceutical options, like the drug aducanumab, with terrible side effects and risks.[5] Boumpas recently stated that evidence supports the use of immunosuppressive medications and glucocorticoids for NP+SLE treatment.[6] In truth, more research has been done on cognitive rehabilitation rather than pharmaceutical treatment, at least with milder cognitive impairments.

Is Treatment Available or Not?

In the not-so-distant past, cognitive deficits (in the general population) were once thought not to be a candidate for rehabilitation since they would likely progress to dementia, but now there is increasing evidence that people with mild deficits can benefit from various cognitive interventions.[7] A "cognitive intervention" is any intervention that focuses on positively impacting a person's cognitive functioning.[7] As an SLP, I would personally recommend traditional cognitive treatment by an SLP or neuropsychologist to help remediate cognitive deficits and to provide support for patients with cognitive deficits. Even the Lupus Foundation of America recommends seeing a cognitive therapist, such as an SLP or psychologist.[8]

Furthermore, a small amount of research is beginning to emerge, which compels scientists to recommend "cognitive rehabilitation" for lupus patients in order to help them "have a better quality of life and to stop the influence of memory impairment on daily activities."[9] Although it is a wonderful recommendation, I am concerned that many lupus patients are completely unaware that cognitive rehabilitation is an option for them. I am also concerned that this recommendation may not be given to them routinely by their rheumatologists.

Surprisingly, in a recent survey of 500 lupus patients, 63% said their rheumatologists had not asked them about their treatment goals in the last three months.[10] Hopefully, patients will advocate for themselves by inquiring about help for their cognitive issues, but their providers should be regularly screening for cognitive issues. An additional issue to overcome is that although individual treatment may be beneficial, it may not be readily available.[2]

Research on Cognitive Treatment in SLE

There are very few peer-reviewed, published research studies on cognitive treatment for lupus patients. Indeed, "the treatment of CI in SLE is exceedingly challenging" since "at this time, there is no clinically proven treatment that effectively targets CI in SLE."[11] However, cognitive treatment is beneficial in other populations, so why wouldn't it work for autoimmune patients? Cognitive treatment is about forming new neural networks or new patterns in the brain, and we now know that the brain remains plastic (capable of learning and changing) throughout life. In one study, seventeen people with SLE received weekly training on memory strategies for eight weeks, resulting in improved memory scores and "metamemory" (strategy use and awareness).[12]

Moreover, in a second study, patients with SLE were found to have significant visual and verbal memory deficits compared to controls, but they were able to improve these areas (specifically with tasks of immediate memory, free & delayed recall, & recognition) with the support of cognitive rehabilitation. The significant improvement prompted the researchers to conclude that "cognitive rehabilitation is considered as a supportive therapy for the neuropsychiatric deficits in SLE patients to have a better quality of life and to stop the influence of memory impairment on daily activities."[9]

Evidence for Cognitive Treatment in General

Although a large body of research does not yet exist for evidence-based treatments specific to SLE patients with cognitive dysfunction, there are standard treatments for other populations suffering from these deficits, and we can borrow from these techniques. In fact, there may be "no specific or unique pattern of cognitive impairment in SLE, but abnormalities include overall cognitive slowing, decreased attention, impaired working memory, and executive dysfunction."[2] In this case, we can extrapolate information gained from evidence-based research in the general population with cognitive dysfunction.

In the general literature, there is substantial research to support treatment for cognitive dysfunction and dementia overall.[13, 14] I have personally seen it benefit many patients in clinical practice. In a systematic review, for example, cognitive training was found to be effective in helping with various aspects of cognitive functions (including processing and executive functions).[14] Moreover, cognitive training may then help autoimmune patients with cognitive dysfunction get back to work, restore self-esteem, and improve their quality of life.[15]

Another systematic review found cognitive intervention to positively affect patients with MCI (mild cognitive impairment) when looking at objective and subjective measures.[13] Furthermore, memory-focused cognitive interventions were found to have a "medium-to-large" positive effect on learning and memory function, as assessed both objectively and subjectively.[16] Similarly, a study found significant positive effects of cognitive intervention on global cognition, executive functions, and delayed memory.[17] It has been concluded that "an appropriately designed intervention can effectively improve memory function, reduce disability progression, and improve mood state in people with cognitive disorders."[16] Furthermore, they found increased positive effects of this memory training with shorter sessions, individualized sessions, and more than eight treatment sessions.[16]

A systematic evidence-based review concluded that several studies support the use of memory strategies and support for people with cognitive deficits.[18] In fact, the researchers found significant gains in ADLs (activities of daily living), memory recall, learning new names, and use of memory strategies. Another systematic review of cognitive interventions in people with dementia found these interventions to be safe and effective for increasing global cognition and quality of life.[19]

Thus, cognitive rehabilitation may improve cognitive functions through various techniques and may also teach strategies to cope with persistent difficulties. In healthy populations, new learning and exposure to rich environments have been shown to promote new neuron survival and incorporation into neural circuits (making these new neurons useful for cognition). Moreover, improved neural circuitry is also what we hope for in cognitive rehabilitation. Unfortunately, there are often constraints with the expense (insurance may not cover cognitive treatment or may cover only a few sessions) and availability. If such services are unavailable in an area (particularly rural areas) or if a patient has difficulty with transportation to or from sessions, one alternative may be telepractice sessions or cognitive treatment delivered remotely through a computer. Although more research is needed, preliminary evidence shows cognitive telerehabilitation may have comparable effects to in-person treatment.[20] Telemedicine exploded during the recent pandemic, of course, and is gaining in favor.

Regarding the general population of older adults with mild cognitive deficits or dementia, the National Academies of Sciences,

Engineering, and Medicine undertook a systematic review of the literature and published a report with three recommendations for treatment: cognitive rehabilitation, high blood pressure management, and physical exercise.[21] Interestingly, they stated that the highest level of evidence from published randomized control trials (low to moderate strength RCT evidence) was for cognitive rehabilitation, although they state additional research is needed in all three areas.[21]

Moreover, a study looked at different forms of cognitive interventions and concluded that therapist-based interventions, as well as multimodal (multiple) interventions, had more impact on ADLs and "metacognitive conditions" when compared to controls and computer-based training.[7] It's important to emphasize the fact that early intervention for patients with cognitive dysfunction may help them maintain cognitive functions and possibly increase certain functions (as in the memory intervention discussed above).[18] According to Masley, "continuous new learning" is the key to protecting your brain and improving cognition since novel learning builds cognitive reserve and creates new pathways between neurons.[22]

Computer-Based Cognitive Training

Computer-based training, in which a patient works on the computer independently, is another area to explore for help with cognitive symptoms. Eighty-four percent of households in the United States own a computer, and computerized training can be more cost-effective than traditional person-to-person treatment.[23] One review concluded that computer-based training has some positive effects, like improvement of short-term memory and learning. However, more longitudinal randomized trials are needed in the research literature.[24] In fact, a robust systematic review based on 17 randomly controlled trial studies concluded computerized cognitive training is a "viable intervention for enhancing cognition in people with MCI."[25]

Not all researchers are in agreement on computerized training, however, as they must consider the transfer or generalization of skills to everyday life activities—not just improved scores. A study, for example, discussed concerns about the limited transfer of these interventions to everyday life.[26] Although aimed at healthy older adults, there are currently several commercially available cognitive programs such as CogMed, Jungle Memory, and Cognifit, as pointed out by research.[24] To that list, I would add BrainHQ (www.BrainHQ.com), which is a program I personally use regularly. BrainHQ has shown improvement in ADLs over time and has been used worldwide.[22]

There is currently a very large randomized-controlled study called PACT (Preventing Alzheimer's with Cognitive Training) enrolling 7,600 adults to determine if BrainHQ will prevent mild cognitive impairment ("$44 Million," 2021). A previous study (ACTIVE study of 2017) found that 18 hours of BrainHQ training decreased the incidence of dementia in older adults by 48 percent ("$44 Million," 2021). In fact, in utilizing BrainHQ with a group of heart patients with mild cognitive impairment, one study found when aerobic exercise was combined with cognitive training, patients had significantly more improvement in verbal memory, which sustained for at least six months, as compared to patients who exercised only.[27]

Another study with healthy older adults found that "speed of processing" computerized visual-perceptual exercises (several hour-long sessions) actually decreased the risk of dementia in those adults ten years later![28] In addition to programs aimed at healthy older adults, there are a number of available computer programs and apps that were designed for people with aphasia or TBI (traumatic brain injury) and are often recommended by SLPs. I believe some of these may be adapted for use with lupus patients with cognitive deficits, depending on which cognitive domains are affected in a particular person.

One recent study utilizing the app HippoCamera to assist those with memory impairment found that individuals who regularly used the app could recall 50% more details of daily events even six months later.[29] As suggested, the effectiveness of computerized cognitive training probably depends on which cognitive domains (attention, memory, executive functions, etc.) are affected in a particular individual, and to what degree.[30] For this reason, I postulate the best approach for people with cognitive dysfunction is an individualized plan, with a combined approach of computerized training and in-person rehabilitation under the direction of an SLP or neuropsychologist.

Cognitive dysfunction in systemic lupus erythematosus (SLE) presents a significant challenge, with treatment options still emerging. While pharmacologic approaches such as cholinesterase inhibitors and agents targeting the blood-brain barrier have shown some promise, they remain inconclusive in addressing SLE-related cognitive impairments. On the other hand, cognitive rehabilitation, once regarded as ineffective for progressive cognitive decline, has gained recognition for its potential in improving memory and quality of life, particularly for patients with mild cognitive impairments.

Evidence supporting cognitive rehabilitation for lupus patients is still developing, but studies have demonstrated its positive impact on memory, attention, and overall cognitive functioning. The role of healthcare professionals, such as speech-language pathologists and neuropsychologists, is crucial in providing the necessary support for these patients. While access to such interventions can be limited, particularly in underserved areas, alternative methods such as telepractice offer hope for increasing availability.

Ultimately, although there is no one-size-fits-all solution for everyone, integrating cognitive rehabilitation into the treatment of SLE-related cognitive dysfunction appears to be the best path forward. More research is needed to refine these approaches and ensure that patients receive comprehensive care tailored to their needs.

Part III:

Treatment and Lifestyle Alternatives

Chapter 11

Will Medications, Biomarkers, or Supplements Help?

As our understanding of cognitive dysfunction in autoimmune diseases like systemic lupus erythematosus (SLE) evolves, so does the exploration of pharmacological and supplemental interventions to alleviate or even prevent cognitive decline. While traditional treatments for mild cognitive impairment (MCI) in the general population have largely failed to demonstrate effectiveness—prompting experts to focus instead on lifestyle interventions such as aerobic exercise and cardiovascular health—research specific to autoimmune conditions presents a more nuanced picture.

In the case of SLE, certain immunosuppressive therapies, corticosteroids, and disease-modifying antirheumatic drugs (DMARDs) show potential in reducing neuroinflammation and preserving cognitive function. However, findings remain mixed and sometimes contradictory. Simultaneously, interest is growing in the potential benefits of neuroprotective supplements, emerging peptide therapies, and repurposed pharmaceuticals. Alongside these treatment efforts, significant research is underway to identify reliable biomarkers—measurable indicators in blood or cerebrospinal fluid—that may signal cognitive decline to identify the need for early treatment in individual patients.

This chapter delves into the current landscape of pharmacologic and supplemental approaches to managing cognitive dysfunction in autoimmune conditions, highlighting both the promise and the limitations of existing evidence while pointing toward a future of more personalized, biomarker-guided care.

Pharmacological Uncertainty

Regarding pharmacology, a literature review for the treatment of MCI in the older general population (not autoimmune patients specifically) found no medications to be proven effective for MCI and suggested treatment focus instead on "aerobic exercise, mental activity, and cardiovascular risk factor control."[1] Similarly, the AAN (American Academy of Neurology) stated no high-quality evidence supports pharmacological treatment for MCI.[2]

Despite these findings in the general population, some research has explored the effects of pharmacological treatments, specifically in patients with autoimmune diseases. One clinical trial claimed to reverse cognitive dysfunction in a woman with SLE by administering corticosteroids and IV cyclophosphamide.[3] However, the evidence for improvement with corticosteroids is mixed, with some studies showing improvement in cognition.[4] At the same time, others show no change or even an increase in cognitive deficits. Some experts believe high doses of prednisone, for example, may be a risk factor for cognitive dysfunction, while immunosuppressive medications may have a protective effect, as mentioned previously.[5] A study stated, "consistent glucocorticoid use, which may be a surrogate of more active or severe disease, is associated with decline in cognitive function."[6]

Additional studies provide further conflicting results. One study found decreased mathematical processing abilities to be associated with daily glucocorticoid use in SLE patients.[7] In contrast, another study showed white matter hyperintensity areas decreased when corticosteroid medication was administered.[8] However, a study found corticosteroid doses to be completely unrelated to SLE patients' brain atrophy.[9] A group of researchers stated that no relationship between corticosteroids and cognitive dysfunction exists.[10] In truth, we don't have many longitudinal studies investigating the long-term cognitive effects of rheumatological drugs.

Pharmaceuticals

Are there other pharmaceuticals that may benefit cognitive impairment in autoimmune disease? One clinical trial tested a different drug - Modafinil - to see if it could improve cognitive dysfunction in lupus.[11] However, the trial did not go forward, so we have to assume the results were not encouraging. Further, a study found increased white matter volume in the SLE patients who were being treated with immunosuppressant therapy as opposed to the SLE patients without this treatment, which, they concluded, suggested a protective role.[12]

After reviewing the existing literature, it was concluded that anti-TNF drugs (a type of biologic DMARD) appear promising in the treatment against cognitive impairment, as opposed to classic DMARDS (disease-modifying antirheumatic drugs).[13] Hydroxychloroquine, the most common drug for lupus, specifically has been suggested to help in the prevention of neuropsychological events, as this type of drug has an antithrombotic effect.[14] However, the vascular effect may apply only to more focal (rather than diffuse) events. Encouragingly, research has shown that patients taking hydroxychloroquine had significantly lower odds of experiencing cognitive impairment.[15] Indeed, antimalarial medications (which many lupus patients take regularly) have been shown to have a protective effect against worsening white matter hyperintensities in the brain.[16]

Interestingly, in one study, simple and consistent use of aspirin was associated with improved cognitive function in SLE patients.[17] Some advocates recommend treatment with antiplatelet or anticoagulation medications if thrombotic processes are present, but with corticosteroids or immunosuppressants if inflammatory processes cause neurological symptoms.[18]

Moreover, the Thymosin beta 4 peptide (currently given by self-injection) has been shown in research to increase tight junction proteins, so it may be a possible pharmacological treatment to improve BBB (blood-brain barrier) stabilization.[19] Low-dose peptides were first found to be successful in murine (mouse) models and are now beginning to be used in humans. One murine study showed peptide treatment could "simultaneously induce Treg cells (which suppresses autoimmunity) and suppress Th17 cells" (causing inflammation), which led to sustained remission of SLE.[20] Therefore, we can assume that they would reduce cognitive impairment. In addition, the peptides suppressed and/or reversed lupus nephritis.[20]

However, when testing treatments of any kind, it is impossible to find a large group of lupus patients with cognitive impairment who are not already on some kind of medication, so there will always be confounding variables.[21] In this regard, a very new study tested the combination of two medications to treat long-COVID brain fog. They combined the use of Guanfacine (originally approved for ADHD) with NAC (N-Acetyl Cysteine), and eight out of 12 patients reported improved memory and/or organization of thought.[22] I would love to see a clinical trial using these two medications with autoimmune patients affected by cognitive dysfunction.

These observations raise several important questions. Would the right medications or supplements, then, have a preventive influence on cognitive function? Could the right combination be the answer for treatment? Alternatively, can some medications contribute to cognitive symptoms, causing them or making them worse? Without further research using medications and supplements, we are left with only questions.

Biomarkers

Research is increasingly focusing on identifying various consistent biomarkers in the blood and CSF of SLE patients with cognitive deficits or other CNS symptoms. As previously mentioned, we don't yet understand how autoantibodies may pass through the BBB or CSF of patients. Nonetheless, several studies have confirmed an important association between antiphospholipid antibodies (aPL) and cerebrovascular disease and psychosis.[14]

In addition, the frequency of Th17 cells has been found to be elevated in the blood of SLE patients.[23] This is a start but does not yet apply directly to cognitive changes in lupus. Expanding on this, some researchers have implicated cytokines, complement components, and autoantibodies in the pathogenesis of NP+SLE.[24] A study showed that increased concentration of NfL (neurofilament light) in CSF was a marker of cognitive impairment in both SLE and Sjogren's patients and was also tied to the presence of anti-NR2 antibodies.[25] Collection of CSF, however, is difficult with lumbar punctures (spinal tap).

However, in the future, we hope to be able to determine the beginning of neuropsychological impairment in lupus by a blood sample long before actual symptoms of cognitive difficulties negatively affect the person's life. We may be getting closer to this goal, "as a significant number of potential biomarkers have been identified to be associated with many clinical aspects of SLE, including diagnosis, disease activity and prognosis."[26]

Associations Between Biomarkers, Health Conditions, and Cognitive Impairment

An important question in lupus research is whether there are specific biomarkers or concomitant diagnoses strongly associated with cognitive dysfunction. Moreover, if such associations exist, it's crucial to determine whether they represent a causal relationship or the result of a common underlying cause. For example, a study involving 694 lupus patients found that "hypertension, stroke, and the presence of aPL (antiphospholipid autoantibodies) were all significantly associated with cognitive impairment."[27]

While it is well established that cognitive impairment commonly follows a stroke with focal damage, and hypertension is related to cognitive impairment due to damaged brain vessels (more diffuse damage), what other health conditions predispose a person to cognitive decline? Can these other diseases/diagnoses offer cues to autoimmune cognitive impairment through similar biomarkers?

Suppose the aPL biomarker is, indeed, frequently associated with cognitive impairment. In that case, researchers need to investigate whether it has a direct effect on neurons leading to cognitive issues or whether the aPL levels cause thrombotic events, etc.[27] In support of this idea, another study found that aPL levels were associated with decreased scores in language abilities and attention to cognitive testing.[28] Both anti-PL (antiphospholipid antibodies) and anti-NMDA (antiblutamate receptor antibodies) are "promising potential biomarkers of CNS involvement."[28]

Serum and CSF Biomarker Research and Future Directions
A group of researchers investigating serum TWEAK as a biomarker of NP+SLE concluded that future research needed to include paired samples of both serum and CSF biomarkers since CSF levels and blood levels often differ.[29] Wang and colleagues did just this, collecting both serum and CSF samples, and found certain cytokines were involved in the pathogenesis of CNS lupus.[30]

Furthermore, elevations of uric acid could also be a potential cognitive function biomarker, as increased uric acid levels were found to increase the risk of cognitive decline in the general population of older adults.[31] However, whether the connection would also hold true for lupus patients remains to be seen. With such specialized information from patient samples, pharmacological interventions may soon be tailored to the patient. A recent article advocates the subdivision of lupus patients into four categories based on various autoantibodies.[32] These types of subgroups may indicate different pathophysiology, leading to more individualized treatment approaches.

Emerging Evidence for Subgroup-Specific Personalized Treatment

Subcategories may be established in the future utilizing molecular/cellular analysis, leading to more personalized treatment decisions.[33] In fact, research recently identified a subgroup of lupus patients with high interferon in the blood.[34] This patient population was made of a higher proportion of South Eastern and North Eastern Asian patients, and they presented with the disease at a younger age. Moreover, autoantibodies were more common in this group; they had a more severe disease pattern affecting multiple domains, and ANA staining patterns were more likely to be speckled.[34] By contrast, the lupus patients with low interferon were more likely to be of Caucasian background, have a more homogenous ANA pattern, and have lower disease activity. The authors concluded this information has "prognostic significance" in managing SLE, as it can guide medication decisions.[34]

Furthermore, researchers think that subgroups can be confirmed and treatment individualized by looking at single cells of lupus patients, utilizing RNA sequencing.[35] They found increased expression of interferon-stimulated genes with individual cells in lupus patients, assessing monocytes, T cells, natural killer cells, dendritic cells, B cells, and plasma cells. By looking at these subpopulations of cells within the blood, they can find subgroups of patients with the highest disease activity and eventually tailor treatment to patients.[35]

Broader Search for Biomarkers and Clinical Implications

Are there easily identified biomarkers in every lupus patient that could direct the physician to the most efficient treatment for that individual? If researchers can identify such specific biomarkers in an individual, this would eventually allow rheumatologists to select a targeted medicine or combination of medicines with an individualized program for each patient.[36] However, "no effective non-invasive biomarkers are available to evaluate the cognitive status of patients." [37] Due to the "multifaceted nature of cognitive dysfunction in SLE," "future therapeutic approaches will need to be individually tailored to address the relevant drivers in individual patients."[21]

In this regard, researchers are currently looking into complement-activating fragments to determine if they can find a good biomarker for lupus.[38] In searching for biomarkers, one study found a "type I interferon signature" in almost 70% of lupus patients.[39] Similarly, another study found that NAA (N-acetyl aspartate), glutamine, and glutamate may be additional biomarkers that indicate cerebral diseases in SLE patients, as they also found changes in the right insula of the brain.[40] Moreover, NAA (found in mitochondria in neurons) was also decreased in the LLN (left lentiform nucleus) of lupus patients in another study.[41]

In parallel, SLE patients with cognitive deficits had significantly higher levels of IGG anti-cardiolipin antibodies.[42] One study found APL (particularly aCl) and anti-dsDNA antibodies elevated in lupus patients with cognitive impairments in attention, visuospatial skills, verbal memory, and executive function.[43] Building on this, researchers are exploring the connection between DNRAb (a subset of anti-dsDNA antibodies) and cognitive dysfunction in SLE, as the DNRAb binds DNA and reacts with NMDAR subunits, the brain's most important receptor.[44] They found DNRAb to be associated with spatial memory deficits.[44]

According to a recent study, Anti-NMDAR (N-methyl-D-aspartate receptor) antibodies and S100Beta proteins are two possible biomarkers.[45] NMDAR is a glutamate receptor subtype that helps in synaptic remodeling. When antibodies attack NMDAR, then that affects mitochondria and calcium, which may lead to apoptosis (death) of neurons and cognitive impairment.[45] These researchers propose that "serum anti-NMDAR antibodies can be used as a predictor for SLE-related cognitive dysfunction.[45]

Moreover, studies have also found the highest levels of anti-NR2 (reactive with NMDA or N-methyl-D-aspartate) in the CSF of lupus patients with "acute confusional state," while lower levels were present with lesser neuropsychological symptoms.[46] Hopefully, with further research, the management of SLE is moving toward a state of "personalized medicine."[47]

A well-known group of researchers calls for the "development of more reliable serum biomarkers for early disease detection" to improve SLE patients' prognosis.[48] Complement C3 may be another possible biomarker of SLE, detected through molecular imaging, as a sensitive marker of inflammation.[49] With more research like this, physicians may soon be able to order specific blood or CSF tests and choose certain treatments based on the results.

Researchers pose several additional candidates for biomarkers that may indicate NP+SLE: autoantibodies, lipocalin 2, osteopontin, albumin, haptoglobin, microglobulins, S100B, and BDNF.[50] A meta-analysis of studies found NP+SLE patients (as compared to SLE patients with no CNS or PNS symptoms) were more likely to have elevated levels of the following antibodies: aCL, LA, APL, anti-ribosomal P Abs, anti-neuronal Abs.[51]

Research linking biomarkers specifically to cognitive issues is just beginning, as scientists discovered an association between raised IL-6 (a pro-inflammatory cytokine) levels and worse cognitive function in SLE patients.[52] Another study found S100B protein levels to be increased in SLE patients with cognitive impairment.[53] In 2022, research found both of the cytokines S100A8/A9 and MMP-9 to be elevated in SLE patients with cognitive impairment.[54]

Recently, one study found an association between cognitive impairment and lupus anticoagulant (LA) and anticardiolipin antibodies (aCL) in the blood of lupus patients.[55] Another group of researchers recently explored Type I Interferon and its relation to serum KYN/TRP(kynurenine/tryptophan) and QA/KA (quinolinic acid/kynurenic acid) ratios as potential biomarkers tied to cognitive dysfunction (or at least working memory) in lupus patients.[56] Comparing 74 lupus patients with matched healthy volunteers, they found a significant association between the imbalance of quinolinic acid & kynurenic acid in the blood and poor cognitive performance in the people with lupus.[56]

Similarly, others have found high levels of tumor necrosis factor (TNF) in the CSF of lupus patients.[14] A very technical immunological article explained how, in people with lupus, B cells (a type of white blood cell) are hyper-responsive to receptor-7 (an immune system protein), which "leads to the generation of autoreactive antibody-secreting plasmablasts."[57] For additional information along these lines, an article reviews several different "drivers" of the immunopathogenesis of SLE, including cytokines, type I interferons, BlyS, plasmablasts, B cells, etc.[58] It is possible that one or more of these markers relating to the pathogenesis of lupus may eventually become biomarkers for cognitive impairment, as well.

Future Directions in Biomarkers and Neuroinflammatory Treatment

Assuming the future holds promise for finding specific biomarkers, allowing for individual personalized treatment, researchers stated, "low disease activity states and prevention of flares are realistic targets in the management of SLE associated with improved prognosis."[59] Similarly, another study stated identifying certain biomarkers would "allow adequate prediction of response-to-therapy" treatment.[60] In summary, subgroups and biomarkers of lupus could increase our knowledge of pathogenesis and inform our diagnostic and treatment procedures.[32] These advancements could lead to more individualized and efficient treatment of cognitive impairment, as well as other symptoms of lupus and other autoimmune diseases.

In addition, another avenue for future treatment may also lie within the cell as structural changes take place. For example, microglia (located in the brain and spinal cord) have been found in mice to play a role in synaptic pruning. Mice were treated with a sphingosine-1-phosphate receptor modulator, which decreased proinflammatory cytokines (secreted by the microglia) and improved memory.[61] Microglia may be overly activated in lupus patients with cognitive impairment, causing synaptic pruning.[50] In fact, in Alzheimer's patients, for example, deposits of amyloid beta plaques in the brain cause over-activation of microglia, which then releases pro-inflammatory cytokines and increases brain inflammation.[13] Microglia can prune dendritic synapses associated with cognitive dysfunction in mouse models when activated.[44] In this way, "there is significant evidence suggesting that microglia play a central role in the inflammatory cascade leading to cognitive impairment in SLE."[62]

Some researchers have proposed finding a medication, like an ACE inhibitor, to suppress microglial activation.[62] This would improve cognitive function. In fact, an exciting trial is underway by these researchers, in which ACEi (angio-tensin-converting enzyme inhibitors) are used to determine if it is effective in improving cognitive deficits in SLE patients. Studies have already shown that ACE (angio-tensin-converting enzyme) inhibitors (captopril & perindopril) have improved cognitive deficits in mice and, thus, might be an effective treatment for cognitive dysfunction in lupus.[50] The ACE inhibitors reduce the microglia activation described above, which would prevent over-pruning of dendritic synapses. Specifically, this type of treatment has been shown to prevent spatial memory impairment in mice.[63] However, in another recent study, the neuropeptide CGRP (Calcitonin gene-related peptide) was found to be effective in inhibiting overactive microglia that were causing neuroinflammation.[64]

In addition to ACE pathways, other inflammatory pathways, such as Bruton's tyrosine kinase, Nogo-a, and TWEAK/Fn14, have all been proposed as potential treatment targets with neuroinflammation.[65] Alternatively, fingolimod, an immunosuppressive drug, has recently been introduced in MS (Multiple Sclerosis) patients with cognitive deficits. In mice, fingolimod was able to reduce memory deficits.[61] So, it might also be a good avenue for research in lupus patients with cognitive impairment. Also, another recent study in mice showed that oxytocin administered intranasally improved memory, and the researchers hypothesized that it may eventually become a useful (and easy) treatment for cognitive impairment in humans.[66]

Supplementation

Supplements may be another area of interest for researchers looking for treatment for autoimmune disease. A suboptimal level of nutrients can be one of the many causal factors for autoimmune disease (as well as decreased cognition) and, thus, an important part of treatment. It is important to note these mild nutritional deficits often do not show up in conventional blood tests but may require testing by a functional medicine practitioner. For example, in mouse models, selenium has been found to increase neurogenesis and reverse cognitive decline.[67] A clinical trial of fish oil supplementation with SLE patients resulted in significant improvement in fatigue levels, although cognition was not formally assessed in the study.[68] Masley states that randomized clinical trials have shown vitamin D supplements help reduce the risk of autoimmune disease (as well as cancer).[69] Moreover, supplements that have been shown to positively affect autoimmune disease include NAC (N-acetyl-cysteine), omega-3, zinc, curcumin, and Coenzyme Q10.[70] Similarly, healthy adults benefitted by taking a nicotinamide mononucleotide (NMN) supplement to increase levels of nicotinamide adenine dinucleotide (NAD+), which directly influences mitochondrial function and may also help with cognition and immune function.[71]

Surprisingly, one study recently found an inverse correlation between caffeine intake and disease activity in SLE, such that patients with a low caffeine intake had more severe disease.[72] Earlier in the book, we discussed the recent theory of lupus, beginning with a disturbed microbiome.[73] If this is the case, it may be possible that a particular combination of probiotics or certain foods could be prescribed to alter the microbiome positively, which may then improve lupus symptoms. Targeted treatment of the microbiome has the potential to be a new avenue for treatment due to recent findings.[74] For instance, investigators found that short-chain fatty acids protect intestinal integrity (or prevent leaky gut).[75] Obviously, more research is needed in this area before specific recommendations can be confidently made.

Some supplements have been found to be beneficial for cognition and improving cognitive deficits in the general population. A good place to start is with a daily multivitamin and mineral supplement, as a recent 3-year study found such supplementation to improve cognitive, memory, and executive function in older adults.[76] Additional supplements are often necessary, however. Low vitamin D levels, for example, are associated with a higher risk of cognitive decline since vitamin D stimulates brain cell growth.[69] In fact, vitamin D levels are often low in SLE patients. A study found a potential relationship between vitamin D levels and cognitive function.[77] Researchers found "patients with SLE had significantly lower vitamin D levels than the healthy controls," and there was a correlation between vitamin D levels and decreased executive functioning.[78] In fact, a study of mice with SLE found supplementing with vitamin D suppressed inflammatory cytokines in the brain, which reduced the number of dead cells in the hippocampus and, thus, improved cognition.[79] Since we know vitamin D is related to sun exposure, we also know 80% of lupus patients are sensitive to sunlight, leading to systemic flares and even kidney disease.[80] I wonder if the low vitamin D levels are directly tied to necessary sun avoidance in lupus patients. (As an aside related to the mystery of how UV radiation makes SLE patients flare, see Grieling et al. (2018) for a possible connection to Ro60 autoantigens.)[81]

Similarly, low levels of vitamin B12 are also dangerous, as this can cause dementia from the death of brain cells.[69] In older adults with "normal" but not "optimal" B12 levels, those with lower levels were found to have slower processing speed and delays in responding to visual stimuli.[82] The researchers concluded, "Ultimately, we need to invest in more research about the underlying biology of B12 insufficiency since it may be a preventable cause of cognitive decline."[82] Moreover, Polyphenon E, obtained from green tea, has been found to have a neuroprotective effect due to the antioxidant EGCG.[83] A recent study found that green tea consumption was associated with less cerebral white matter lesions and concluded that drinking 3+ glasses daily may help prevent dementia.[84] Similarly, Black tea is also extremely beneficial, as its catechins and theaflavins have anti-inflammatory and antioxidant effects.[85]

For cognitive health, Dr. Masley recommends finding a very good multivitamin and supplementing it with vitamin D, long-chain omega 3's (fish oil), and a probiotic.[69] There is also some evidence that the spice turmeric has a neuroprotective effect, as cultures that eat large quantities of it have much lower rates of dementia. Curcumin (found in turmeric) increases BDNF (the brain-derived neurotrophic factor that helps grow neurons) and helps dendrites (branches of neurons) grow, at least in mice.[86] In addition, resveratrol has been found to improve neurocognitive impairment (associated with ischemia) in rats.[87] Resveratrol, found in red grapes and dark chocolate, is both neuroprotective and anti-inflammatory.[88]

Furthermore, a review of the literature found definitive evidence for supplements or dietary factors benefitting cognition, particularly for B vitamins (especially folic acid), DHA (omega 3), and flavonols.[89] DHA supplementation has even reversed memory deficits in rats.[90] Serum fatty acid profiles were recently analyzed in SLE patients, and better outcomes were found with increased tissue levels of omega-3s, particularly DHA.[91] Vitamins A, C, and E help improve oxidative stress associated with cognitive decline.[89] In addition, ginger has anti-inflammatory and anti-oxidative effects and was found to have a protective role in mice with lupus by improving autoantibody formation and vein thrombosis.[92] Magnesium threonate, which crosses the blood-brain barrier, may have neuroprotective effects.[93] Choline is a vital nutrient that needs to be obtained in food (meat, poultry, beans, broccoli, eggs) or supplements for brain health since it is necessary for the production of acetylcholine, which affects memory.[94]

A recent study reviewed the literature (research studies) to address questions frequently asked by lupus patients about diet and supplementation.[95] They composed seven statements or conclusions based on what they found. One said, "limited evidence suggests that supplements like vitamin D, curcumin, and omega-3 fatty acids may influence immunological inflammatory markers." Another statement was that "some studies suggest that vitamin D and to a lesser extent, vitamin E supplements, might play a role in managing immunological inflammatory parameters." They concluded that there is a need for further investigation into nutrition and lupus.[95]

You may recall a discussion of cellular senescence or cellular aging in the chapter on the etiology of lupus. Some researchers think lupus is a sort of rapid aging, with immunosenescence, or aging of the immune system, beginning very early in the patient and contributing to MCI, among other symptoms. Interestingly, natural senolytic agents or supplements/food nutrients that fight such premature cellular aging have been identified. They include: curcumin, piperlongumine, fisetin, and quercetin.[96]

In conclusion, the use of pharmacological treatment and supplementation for cognitive dysfunction in autoimmune disease remains an area of active research with conflicting results. While studies suggest that certain medications, such as corticosteroids and immunosuppressants, may improve cognition in some patients, the evidence is not conclusive. High-dose corticosteroids, for example, are often associated with cognitive decline, whereas medications like hydroxychloroquine and anti-TNF drugs appear to have a more protective effect. Recent findings also indicate that peptides like Thymosin beta 4 may stabilize the blood-brain barrier and offer potential therapeutic benefits. Regarding supplementation, various vitamins, nutrients, and spices are tied to improved cognition in the literature, and may be worth exploring further in autoimmune disease.

Additionally, biomarkers hold promise for better understanding cognitive dysfunction in lupus and for developing personalized treatment approaches. Research into autoantibodies, cytokines, and other biomarkers may help identify subgroups of patients who would benefit from specific treatments. The development of personalized medicine based on these biomarkers could lead to more effective management of cognitive impairment in lupus patients.

While progress is being made, much remains to be explored. Future studies are needed to clarify the role of different medications, supplements, and biomarkers in managing cognitive dysfunction. The goal should be to find targeted therapies that improve cognitive function, reduce disease activity, and prevent further damage to the central nervous system in patients with autoimmune diseases like lupus.

Chapter 12

What Else Can We Do To Improve?

Cognitive dysfunction is a common challenge faced by many individuals, including those with autoimmune conditions like lupus. While limited research specifically addresses cognitive issues in patients with systemic lupus erythematosus (SLE), valuable insights can be drawn from broader studies on cognitive impairment across various patient populations. As the aging population grows, so does the prevalence of cognitive decline, leading researchers to investigate effective treatments.

This chapter covers the impact of exercise, sleep, and supportive aids, as well as innovative therapies, on cognition. By examining the latest research, we uncover strategies that can help improve memory, attention, and overall cognitive function. Despite the challenges of implementing a few of these interventions, there are practical, accessible approaches that can begin to make a difference today.

Exercise and Cognition

Since we have very little research guiding us on how to help SLE patients with cognitive dysfunction specifically, we must broaden our scope to look across patient populations and see what research is revealing about general treatment for cognitive deficits. With the dramatic rise of the aging population, cognitive deficits and dementia have significantly increased within the general population.[1] Studies consistently find spending more time doing physical activity and less time in sedentary behavior is associated with better cognitive skills.[2] Further, many investigations of the general aging population have examined the relationship between exercise and cognition. Studies have demonstrated positive effects of physical activity (exercise) on adult neurogenesis and improved memory, attention, and executive function skills.[3] Animal studies have even shown that physical exercise produces adult neurogenesis (growth of new neurons).[4] Animal models have also determined that physical exercise is synaptogenic (forms synapses between neurons), and this has been shown recently with 400 hundred older adults.[5]

In line with this, a 2018 study found fitness to be the strongest predictor of cognitive function.[6] A recent review article identified three types of exercise beneficial to cognition: aerobic, anaerobic (like high-intensity interval), and resistance exercise.[7] Another article found cardio or aerobic exercise to be the most important type for decreasing all-cause mortality, but adults who added weight training to their routines had additional benefits.[8] Researchers are finding exercise is an important tool to counteract cognitive decline, and there may be an exercise "sweet spot," in which the specific duration of exercise matters in terms of improved hippocampal (memory) function.[9]

Many lupus patients, however, find exercise extremely difficult, as fatigue and pain are two prominent symptoms of the disease. In a survey, 92% of SLE patients believed exercise was beneficial, but over 78% said lupus "impeded their ability to exercise."[10] Interestingly, studies have shown the "aerobic capacity of patients with mild SLE is comparable to that observed in patients with severe cardiopulmonary disease."[11] One study examined the relationship between SLE, cognition, exercise, and obesity and found obesity and physical inactivity were independently associated with cognitive dysfunction.[12] While I can certainly relate to the fatigue of lupus, beginning with small, gentle movements can make a difference. With the help of a health coach or trainer, patients can gradually grow exercise habits at a pace that is comfortable for them.

Similarly, another study of 138 lupus patients found only 5% of those who exercised had any cognitive dysfunction, while 23% of those who didn't exercise had memory deficits.[13] What is encouraging is that walking for exercise only three times a week was found to be enough to improve episodic memory for adults over 55.[14] If you're able, counting steps could be a goal since 8-10,000 per day has been shown to increase longevity for adults under 60, while adults over 60 only need 6-8,000.[15] An additional study found walking 9,800 steps per day to be the optimal dose to prevent dementia.[16]

Although there are challenges in finding such a program, the possibility of combining thinking with physical exercise shows promising results. A meta-analysis of using exergames (exercise thinking games) to improve cognitive functioning revealed significant improvement in executive functions and visuospatial perception but not in attention or overall general cognition.[17] A clinical trial involving cybercycling (stationary bikes with virtual reality tours) concluded that cybercycling (vs. traditional exercise) caused a greater improvement in executive functions and even achieved enhanced neuroplasticity.[18] In a group of stroke survivors, one randomized control trial found that their cognition improved more with a combination of thirty minutes of aerobic exercise and thirty minutes of computerized cognitive training.[19] A study found **both** cognitive intervention and physical exercise improved global cognitive function.[20] In addition, the combination of these treatments appears to impact ADLs positively.[21]

These studies show that cognitive training combined with exercise and cognitive training immediately following exercise are good options. Interestingly, the AAN (American Academy of Neurology) tells clinicians to recommend regular exercise and that they *may* also recommend cognitive training for patients with MCI.[22] In addition, there are numerous additional health benefits, as exercise has been shown to reduce levels of fatigue and depression in lupus and improve cardiorespiratory factors.[23]

Sleep and Cognition

Sleep is another vital factor in improved cognition. Getting the right amount of sleep is important, as getting too little or too much is associated with decreased cognition.[2] Brain changes that alter plasticity, as measured by transcranial magnetic stimulation, correlate with impaired learning, memory, and attention in sleep-deprived individuals compared to those with sufficient sleep.[24] Sleep quality is important, as well, as it is not only about the time you spend in bed. If you are waking up exhausted and foggy, it may be time to request a professional sleep evaluation. It is important to make sure that oxygen levels are not dropping too low at night. If a formal evaluation isn't a possibility, there are many options for sleep trackers now, including apps, Oura rings, Apple watches, etc.

Moreover, certain supplements can be helpful if you are having trouble getting enough quality sleep. For this purpose, talk to your doctor about considering melatonin or glycine at bedtime. Glycine has been found to improve sleep quality in some patients.[25] Apart from Glycine, melatonin is a well-known sleep aid and has also been found to improve cognitive testing scores in patients with Alzheimer's disease.[26] Please stay away from prescription sleep medications, as they have been associated with dementia.[27] Additional tips for improving sleep can be explored with a health coach, such as removing screens/blue light 1-2 hours before bed, and having a predictable, relaxing nightly routine.

Supportive Aids

Supportive technology and external aids are additional areas in need of research. In other populations with cognitive dysfunction, these supports have been found to be beneficial for some patients. People with cognitive deficits may find help with memory deficits, time orientation, and even safety by utilizing smartphones with digital calendars, to-do lists, GPS (Global Positioning System), etc.[28] These things can be part of common-sense daily organization techniques, such as those proposed by Griffin: "write everything down; prioritize your tasks; say things out loud; work at your best daily time"; etc.[29] In addition, external aides like pill boxes and voice recorders can be helpful for memory deficits.[30] There are books on memory aids, if you're interested, such as Jordan's (2020) *Coping with Mild Cognitive Impairment*.[31]

Borrowing again from research based on other populations, recent recommendations for memory impairment with TBI (traumatic brain injury) patients include using a combination of group/individual rehabilitation (led by a clinician) and the use of either smartphone reminders or paper/pencil organization, depending on the preference of the patient.[32] In the research literature on utilizing supportive technology with older adults with MCI or dementia, it has been pointed out that few studies report on how supportive technology influences the quality of life, employment, or human dignity.[28] They further state, "human dignity is closely related to human identity."[28]

Innovative Options

A novel treatment for Alzheimer's Disease (AD), which, if validated, might someday be utilized in patients with milder cognitive deficits, is currently being explored by "Cognito Therapeutics." The Cognito therapy stems from research with mouse models, which showed that amyloid plaques in mice with AD could be reduced in the prefrontal cortex and hippocampus (thinking and memory centers) simply by playing clicking noises and light flashes at 40Hz for an hour a day.[33] Since patients with AD are thought to have hyperactive neurons, the treatment is hypothesized to "reset" the brains of these mice. An even more recent study tested audiovisual 40 Hz light and sound one hour a day in human patients with amnesic MCI or prodromal Alzheimer's. They found success in strengthening functional connectivity and increasing gamma neural activity and stated that this exciting and non-invasive treatment should be further investigated in patients with cognitive impairment.[34] Researchers caution that such standard use of audiovisual therapy with humans is years away. While it might seem strange to utilize Alzheimer's research in a book on autoimmune disease, there are some definite connections. Recent Alzheimer's research has shown a systemic immune response, including pro-inflammatory cytokines & increased activated T cells.[35] Some researchers even think Alzheimer's will soon be declared an autoimmune disease. For example, Weaver wrote an article in the journal *Current Alzheimer Research* entitled, "Beta-Amyloid is an Immunopeptide and Alzheimer's is an Autoimmune Disease."[36]

Another technique, which has been done only with mouse models so far, appears simple. It involves giving an oxytocin peptide intranasally. Researchers found this method to be effective in improving cognition in mice with Alzheimer's disease.[37] Wouldn't it be wonderful if improving cognition was as simple as a nasal spray?

Another new encouraging treatment is light therapy, which, in the testing phase, used lamps with filters that only transmit UV-A1 wavelengths. The light treatment for patients with lupus "significantly decreased disease activity, inducing an early reversal of fatigue, depression, and cognition dysfunction."[38] This treatment seems counterintuitive, as it is well known that most lupus patients are photosensitive to sunlight, but the specific treatment technique isolates a UV wavelength that has been found to be healing in both mice and humans with lupus, actually reversing some of the damage from sunlight. Other studies have also found improvement with UVA-1 phototherapy in lupus patients, particularly with fatigue.[39]

Another new area of research is brain entrainment. In these studies, researchers used EEG machines and found that individuals learn faster when learning in sync with their own brain rhythm.[40] A study using EEG biofeedback with cancer patients reported alleviation of cognitive impairments associated with the "chemo brain."[41] Related to this is targeted transcranial neuromodulation. Researchers discovered that through entrainment of theta and gamma rhythms, working memory and long-term memory could be improved in older adults, with those with the lowest initial cognition (tested by the MoCA) making the biggest gains.[42] Further research in this area may enhance cognitive rehabilitation options for autoimmune patients.

Some studies are beginning to find results with hyperbaric oxygen therapy. Oxygen treatment has increased cerebral blood flow and improved memory in Alzheimer's and dementia.[43] "Increased CBF(cerebral blood flow) and higher oxygen delivery can improve or boost brain function."[43] Hyperbaric oxygen treatment has been proposed as a treatment for mild cognitive impairment, as well as Alzheimer's disease.[44] Perhaps this type of treatment could be explored in autoimmune diseases as well.

Another interesting research study on autoimmune disease is the possibility of stem cell transplants reversing symptoms. For example, a group of researchers discovered symptoms of multiple sclerosis (another autoimmune disease) were reversed when "nanosized exosomes from bone marrow stem cells" were injected into mice with the disease.[45] This type of treatment began to be tested in human trials in 2020, hoping that success will have positive implications for patients with all kinds of autoimmune diseases. Sometimes, lupus patients can achieve remission with stem cell treatment due to the generation of Treg cells, although the same positive result may be able to be achieved with peptide treatment or, possibly, even with probiotics.[46, 47]

Although the above techniques may sound peculiar or unfamiliar, it is important to explore all options to relieve the suffering caused by cognitive disability. Importantly, until such treatment becomes mainstream, we already have techniques under our control, such as those mentioned: exercise, sleep, cognitive treatment, and cognitive support aids. Do not discount the value of such habits you can begin to implement them today. For more information and to explore the additional, highly important strategy of optimal nutrition, please read chapters fourteen and sixteen.

* * *

In conclusion, addressing cognitive dysfunction in autoimmune diseases like lupus requires a multifaceted approach that incorporates exercise, sleep management, supportive aids, and innovative therapies. Research highlights the benefits of physical activity, including improved neurogenesis and cognitive function, although challenges such as fatigue and pain in lupus patients need to be considered. Strategies such as cognitive training combined with exercise show promise for improving memory and executive function. Sleep, too, plays a crucial role in cognitive health, with quality sleep supporting memory, attention, and learning. For those struggling with sleep, supplements such as melatonin and glycine may be discussed with medical practitioners.

Supportive aids, such as digital reminders and organizational tools, can help manage memory deficits, while emerging treatments like light therapy, brain entrainment, and hyperbaric oxygen therapy offer exciting possibilities. Although these therapies are still in early stages, they provide hope for more effective treatments in the future. Until these novel therapies become widely accessible, it is essential to focus on accessible and proven interventions, such as regular exercise, cognitive training, and proper sleep. By integrating these strategies into daily routines, lupus patients can take proactive steps toward improving their cognitive health and overall well-being.

Chapter 13

How About Quality of Life & Nutrition?

Living with lupus involves more than managing physical symptoms; it affects every aspect of a person's daily experience, from emotional well-being to the ability to think clearly and carry out simple tasks. In this chapter, we explore how lupus impacts quality of life (QOL), particularly through its effects on cognitive function.

The emotional toll of cognitive dysfunction, commonly referred to as "brain fog," can be profound, affecting self-identity, relationships, and independence. But alongside this challenge, we also examine a powerful and often overlooked ally in improving both QOL and cognitive health: nutrition. Emerging research and personal experience suggest that what we eat can directly influence how we think, feel, and function. This chapter brings together scientific evidence and patient insights to consider how nutritional strategies may support better outcomes for those living with lupus.

Quality of Life

Further discussion is necessary regarding quality of life (QOL), which is significantly negatively affected by lupus, as mentioned at the beginning of the book when we defined lupus. "Quality of life" was defined by the World Health Organization in 1990 as "encompassing personal goals, expectations, standards, and concerns related to living conditions, including physical and mental functions, role activities, social adaptability and the overall feeling of health."[1] Keeping this in mind, a recent study assessed the real-world burden of lupus and found the diagnosis associated with economic burden, poorer productivity, and decreased QOL, citing the need for improved therapeutic options and increased patient support.[2] Further, researchers have found that psychological intervention can improve certain aspects of QOL and enhance some physiological functions.[1] The Outcome Measures in Rheumatology international consensus effort recommended, in 1998, five domains of assessment in all studies of lupus, and quality of life was one of these domains.[3]

Lupus patients with cognitive deficits tend to have a further decreased QOL. As stated by a group of researchers, "cognitive dysfunction is common in SLE and significantly impacts quality of life."[4] Specifically, cognitive impairments, although considered a diffuse symptom of CNS disturbance (as compared to a focal deficit like a stroke), have "debilitating consequences for patients."[5]

A study on neuropsychiatric symptoms in systemic lupus found QOL to be significantly reduced in lupus patients with neuropsychiatric symptoms as compared to patients with other chronic diseases (like migraine).[6] They defined QOL as "the functional effect of an illness and its consequent therapy upon a patient, as perceived by the patient" and found QOL to be decreased for lupus patients in both physical and mental domains.[6] We must not forget that cognitive deficits can negatively affect a patient's adherence to medical treatment, which is one aspect of QOL.[7] In addition, reduced functional cognition has been found to reduce overall participation in daily activities.[8] One survey of patients with lupus found forgetfulness and lack of concentration to be "causing substantial harm to their lives," although they usually referred to this as general "brain fog."[9] Interestingly, cognitive stimulation has been found to improve the quality of life in dementia patients.[10] Might cognitive stimulation be part of the solution for lupus patients? The numerous ways lupus attacks the body are varied and painful, changing year to year, but the way it attacks the brain is highly destructive to a person's identity.

Cognitive dysfunction becomes a private battle with one's own mind.

Personally, I am aware of how QOL impacts everyday life and why cognitive dysfunction is so devastating. Having a cognitive disorder makes you feel as if you are losing your identity and your dignity. You no longer function as the person you know you are. It is embarrassing and frightening when you can no longer trust your thinking. You begin to question who you are and your ability to work and care for your family.

Introduction to Nutrition

Does nutrition play a role in cognitive changes and, thus, quality of life? I believe it is highly important based on everything I have read and experienced personally. Recent research has made the connection between healthy eating and better cognitive performance. Diet is tied to inflammation and cognitive health.[11] African-American women with high carbohydrate and low-fat diets were found to have a significantly increased risk of SLE/lupus.[12] A ground-breaking new study found a higher daily intake of UPF (ultra-processed foods) was associated with an over 50% greater risk of lupus/SLE.[13] Even without this knowledge, many patients with SLE and other autoimmune diseases have discovered that they must follow specific diets (gluten-free, dairy-free, nightshade-free, low sugar, etc.) to help keep symptoms under control. Ceccarelli et al. (2018) found a significant association between high cholesterol and cognitive impairment (except for attention).[14] It was in agreement with a previous study involving nearly 1,000 SLE patients.[15] As discussed in the Epilogue at the end of the book, I personally saw an almost immediate improvement in my cognitive deficits following significant dietary changes.

Could cognitive changes, at least in part, be tied to insulin resistance? Insulin resistance and diabetes (directly affected by diet/nutrition) have been repeatedly connected to poorer brain function. This occurs due to insulin resistance causing brain cells to obtain less energy through mitochondria (energy powerhouses) malfunction and producing abnormal free radicals and increasing inflammation with oxidative stress. Interestingly, some researchers have begun to refer to Alzheimer's disease as Type 3 diabetes since they believe the cognitive deficits are directly related to the brain's inability to efficiently utilize glucose.[16] Masley points out the importance of differentiating between what foods we eat and what nutrients we are getting (whether through food or supplementation), as certain nutrients are critical for brain health.[16]

Two recent studies (one in humans and one in rats) found increased consumption of nuts (particularly walnuts and peanuts) to be linked with improved cognition.[17] Similarly, another study found that increased mushroom intake (over two portions per week) reduced the odds of having mild cognitive impairment in older adults.[18] Related to this, certain medicinal mushrooms (or their extracts) have been shown to help cognition and reduce neurodegenerative diseases.[19, 20] See chapter 15 for more detailed diet and nutrition information. In addition, there is an emerging science in the field of nutrigenomics that explores the variations among individuals and what nutrition they need based on their genetic makeup. More research is needed, however, before we understand the connection between a specific diet, nutrients, and cognitive health.

In conclusion, living with lupus presents challenges far beyond the physical symptoms, as cognitive dysfunction—often termed "brain fog"—greatly impacts an individual's quality of life. Cognitive impairments in lupus patients have been linked to decreased emotional well-being, social interactions, and daily functioning, further exacerbating the illness's burden. Importantly, emerging research highlights the crucial role nutrition plays in managing these cognitive symptoms. A balanced diet can help mitigate inflammation and improve cognitive performance, showing promise in enhancing quality of life for lupus patients.

Studies have demonstrated the connection between dietary patterns, insulin resistance, and cognitive health, indicating that factors such as nutrient intake, cholesterol levels, and ultra-processed foods may significantly influence brain function. While more research is needed, the current evidence suggests that dietary modifications—such as increasing the intake of walnuts, mushrooms, and other brain-boosting foods—may hold great potential in improving cognitive health and overall well-being for lupus patients. By understanding the impact of nutrition on cognitive function, individuals with lupus can gain better control over their health and find a path toward greater stability and clarity in their lives.

Chapter 14

Will It Progress & Can We Prevent It?

Cognitive dysfunction in lupus presents a complex and often unpredictable journey. Unlike many conditions where cognitive decline follows a more linear or progressive pattern, research in systemic lupus erythematosus (SLE) reveals a different narrative— one marked by inconsistent patterns. This chapter will take you through what is currently known about the progression of cognitive impairment in lupus, uncovering studies that challenge long-held assumptions and offer hope for reversibility. We also turn our attention to prevention, examining emerging insights that suggest the potential for early interventions and lifestyle factors to reduce risk. While definitive answers remain elusive, growing evidence points toward a future where cognitive health in lupus may be better understood, monitored, and preserved through individualized approaches.

Progression of Cognitive Impairment

What about the progression of cognitive dysfunction in lupus? Cognitive function may change over time, and these "cognitive function trajectories" have not been studied nearly enough.[1] Is cognitive impairment reserved only for those with advanced disease? Apparently not, since studies have found cerebral volume (gray and white matter) can be reduced in SLE patients (driven by inflammation) very early after diagnosis but can improve later as the disease improves.[2] The most hopeful research that I am excited to share with you is a study by a group of researchers in which they initially tested the cognition of lupus patients and then again after ten years.[3] After 10 years, cognition had improved in 50% of patients, while only 10% had poorer scores. The patients' test results revealed a trend in improvement in all cognitive areas, and their executive function performance was significantly better.[3] Similarly, it has been found that lupus patients with longer SLE duration were less likely to have cognitive impairment.[4] Another study compared lupus patients who were younger in age to lupus patients who were older and found that younger patients were more significantly affected by cognitive dysfunction than older SLE patients.[5] Researchers propose that cognitive deficits are reversible.[6] A study states, "individual subjects oscillate between being impaired and unimpaired at different points in time," and "fluctuation in cognitive function in SLE patients appears to be common."[7] Research states that "although sometimes profound in individual cases, the majority of patients have subtle and frequently subclinical cognitive deficits which are evanescent rather than progressive over time."[8] If deficits are not progressive in most, does this mean that we can just sit back and wait for everything to clear? No, because not all cognitive decline automatically reverses in autoimmune patients, and all people (regardless of disease) must work to stay cognitively sharp as they age.

In contrast to these studies of change over time, for example, one study of 51 patients showed persistence of cognitive deficits but no progression after one year.[9] One study found older age (over 60) was not associated with increased cognitive dysfunction in lupus patients, but not all scientists agree.[10] Other experts say "lupus fog" (aka "brain fog") doesn't usually get progressively worse but gets better and worse over time.[11] Additional studies have also found cognitive decline is not typically progressive in lupus.[8] In other words, cognitive function is variable. I have experienced this variability of cognition daily, with no way to predict cognitive abilities on a certain day. Monitoring cognitive dysfunction over time (progression, resolution, or improvement) has not been well-studied and requires further research.[12, 13] More longitudinal studies are needed to determine if cognitive dysfunction and brain injury are continuous or not and if biomarkers such as increased S100B levels may help track cognitive impairment.[14] As stated in a previous chapter, there is a need for research to find "standardized and validated biomarkers to be used in clinical practice."[15] Experts are not yet sure if neurological injury (and associated cognitive deficit) is transient or permanent.[16] It has also been stated that one of the problems with testing for cognitive impairment in lupus is that "symptoms are often fluctuating and sometimes evanescent."[17] When cerebral atrophy is identified and associated with cognitive dysfunction, researchers do not yet understand the progression of that cerebral atrophy, if any.[18]

We need further studies to determine if cognitive impairment in lupus can be reversed, but one group of researchers has completed early research indicating a possible reversal of cognitive deficits utilizing ACE inhibitors to stop overactive microglia.[19] Again, these results need to be verified and expanded. I agree with the researchers, who stated, "the ultimate goal is to design appropriate, personalized therapeutic strategies" since everyone is different.[20] Moreover, "Cognitive dysfunction in SLE means any therapeutic interventions should be individually tailored."[21] Hopefully, these individualized programs will slowly become a reality, as some research shows positive alterations in neuroimaging following specific cognitive intervention programs.[13] "Future interventional studies will need to stratify patients for more individualized treatment approaches."[22] Individuals with lupus are suffering **now**, though, and they need personalized treatment programs to treat cognitive issues **today** to improve their daily lives.

A Word on Prevention

What about prevention? Can we do anything to keep cognitive dysfunction from happening to lupus patients? Many researchers say we won't know how to prevent cognitive impairment in lupus patients until we understand what is causing it and how we can help, but there are definite lifestyle changes that have been shown to prevent cognitive decline. A study found higher education to be inversely correlated with cognitive dysfunction.[4] They found that lupus patients with more education were less likely to suffer from cognitive deficits. They hypothesized this had to do with increased "cognitive reserve."[4] In the general literature about cognitive changes in older adults, research has mostly focused on understanding and treating cognitive deficits after they have developed, instead of focusing on prevention.[23] Masley discusses a phase in the progression of Alzheimer's disease called "subjective cognitive impairment" (SCI), which precedes MCI.[24] I'm sure there are similar phases within autoimmune cognitive impairment that, when identified early, can be treated.

We do not yet have the research results to know for sure if something like this (a milder decrease in mental functions) might exist in autoimmune disorders and, thus, might be tracked and treated prior to the development of more significant cognitive deficits or dementia. A recent study revealed that 10% of adults age 65 in America have dementia, while another 22% have MCI.[25] With such a high prevalence of cognitive disorders, one would think prevention would be a higher priority in our society. Since we do know from research that brain changes may occur years before cognitive symptoms appear, prevention is important very early in life. Please see Chapter 15 and read Bredesen's book titled *The End of Alzheimer's Program: The First Protocol to Enhance Cognition and Reverse Decline at Any Age* regarding various lifestyle factors that can prevent and reduce cognitive decline. I have personally seen this program work, and I'm now applying these principles when coaching those with autoimmune cognitive impairment.

✶ ✶ ✶

In conclusion, the progression of cognitive dysfunction in lupus presents a complex and unpredictable pattern, with research revealing both fluctuating symptoms and, in some cases, improvements over time. Unlike many other conditions, cognitive decline in lupus does not always follow a linear trajectory, and studies have shown that cognitive function may improve for many patients over the course of years. However, the variability of cognitive symptoms—"the fog"—poses a significant challenge for both diagnosis and treatment.

While some studies suggest that cognitive impairment may be reversible, further research is needed to better understand how and why these changes occur and whether specific treatments, such as ACE inhibitors, can help mitigate the damage. Prevention also remains a key focus, with studies indicating that early interventions, lifestyle factors, and increased cognitive reserve may reduce the risk of cognitive decline. As research continues, personalized, individualized treatment approaches will be critical to managing cognitive dysfunction in lupus patients, offering hope for more effective interventions that can enhance quality of life and preserve cognitive health in the future.

Chapter 15

What Can You Do at Home?

Understanding the brain's health goes far beyond memory games and mental exercises; it encompasses a holistic view of how our lifestyle, nutrition, environment, and even gut health play a role in cognitive function. In this chapter, we'll explore how brain-derived neurotrophic factor (BDNF) encourages neuronal growth and oxygen delivery to the brain, why starting brain-healthy habits early matters, and the growing body of research that links brain health to stress reduction, detoxification, and lifestyle adjustments.

Drawing from key insights in the work of Dale Bredesen and Tom O'Bryan, we'll uncover practical, science-backed strategies for improving and protecting brain function, especially for those dealing with autoimmune conditions or cognitive concerns.

Hope Found in Multiple Factors

In reviewing all the ideas about treatment and prevention, it seems apparent that a simple one-treatment fix or one-change prevention is not the answer. Instead, we must be open to a combination of approaches in both treatment and prevention. Conventional medicine (or the type of medicine we are most familiar with in America) has no real recommendations for mild cognitive impairment (MCI) or brain fog, so we must instead turn our attention to functional medicine and integrative treatment, which incorporate lifestyle changes.

In this regard, a group of researchers stated that "there were no studies that measured the effectiveness of lifestyle and/or cognitive strategies in combination with their dietary intervention."[1] Thankfully, this is no longer true, as we now have significant research tying lifestyle modifications to better cognition.[1] For example, a 2015 study showed improved cognitive function and slower cognitive decline in their multi-therapy intervention group of older adults at risk, which included a certain diet, exercise, cognitive training, and risk factor reduction.[2] Similarly, another study found a combination of cognitive training and social interaction was beneficial to patients with dementia.[3]

In addition, Dr. Dale Bredesen, an innovative neurologist, has developed a specific, comprehensive program based on research, which has been successful with early Alzheimer's disease that examines 30+ factors and addresses 7 key areas (nutrition, exercise, sleep, stress, stimulation, detox, supplements).[4] In a recent study of twenty-five patients with mild cognitive impairment, Dr. Bredesen's "precision medicine approach" was successful in showing improvement as measured by more than one index.[5] While lupus is certainly not Alzheimer's, we can learn from scientists' research about lifestyle changes that benefit cognition. Dr. Steven Masley (2018), for example, advocates a four-pillar approach to a better brain, with the four pillars as follows: 1) food, 2) nutrients, 3) exercise, and 4) stress management.[6] By taking a wide lens view of the research literature, we can learn from research showing success with cognitive improvement and preventing cognitive decline in other target populations.

Thomas's Approach
In "The Lupus Encyclopedia," Thomas lists several ideas for patients to enhance their cognitive success at home in addition to those above.[7] First, he suggests keeping your mind active by working (if possible) or volunteering. Second, you should train your brain with crossword puzzles, Scrabble, etc. (I would extend this further with the BrainHQ program mentioned earlier in chapter ten.) Third, read and study books on memory & cognition

strategies. Fourth, challenge your brain with a new hobby or language.

In addition, Thomas also recommends compensatory strategies, such as keeping a daily calendar, to-do list, and notes.[7] Common sense suggestions include eliminating unnecessary distractions, focusing on one thing at a time, and repeating things you must remember.[7] I would suggest reading books such as *The Lupus Encyclopedia: A Comprehensive Guide for Patients and Families* by Donald E. Thomas (2014), *The Better Brain Solution: How to Start Now at Any Age and Prevent Insulin Resistance of the Brain, Sharpen Cognitive Function, and Avoid Memory Loss* by Steven M. Masley (2018), and *The End of Alzheimer's Program: The First Protocol to Enhance Cognition and Reverse Decline at Any Age* by Dale E. Bredesen (2020). You might also find Dr. Bredesen's earlier work, *Reversal of Cognitive Decline: A Novel Therapeutic Program (2014)*, highly valuable. Engaging with these resources can help stimulate your cognitive thinking and increase your awareness of how subtle cognitive changes can occur over time. As Dr. Bredesen reminds us, "cognitive decline often sneaks up on you," so be vigilant and aware.[4]

Bredesen's Approach

In 2020, Dr. Bredesen published *The End of Alzheimer's Program: The First Protocol to Enhance Cognition and Reverse Decline at Any Age.* I highly recommend his book to anyone with an autoimmune disease experiencing cognitive changes. Again, the book/protocol is not specific to autoimmune disease, but autoimmune disease is mentioned in it, and it is written to apply to the prevention and treatment of any cognitive impairment or dementia. The overall treatment takes place in three phases: removal (of toxins/disease processes), resilience (by doing everything you can to make yourself healthy), and rebuilding (enhancing cognition with new learning). There are many factors to consider within each of these categories presented in the book that people with cognitive concerns (literally everyone) can consider for their own benefit.

As a personal example of my case, I needed the removal/addressing of bad bacterial overgrowth in my gut (dysbiosis) and the decrease of overall inflammation and healing of my intestinal hyperpermeability (leaky gut). Resilience included drastically changing my diet (to address food sensitivities and decrease inflammation), taking supplements (for nutritional deficiencies), exercising almost daily, and getting plenty of sleep. The rebuilding phase (continual) includes challenging my brain with exercise and activity (both physical exercise and brain/thinking exercises).

Dr. Bredesen points out that atrophy of the brain on MRI often does not appear until long after cognitive decline starts, so one should start this protocol as early as possible, either to prevent cognitive decline around middle age or, most definitely, as soon as any cognitive changes are noticed.[4] He goes into great detail about a specific diet for cognition and emphasizes the importance of ketosis and insulin sensitivity. In this way, diet has been found to be a powerful tool to decrease systemic inflammation.[8] One big part of Bredesen's diet is the elimination of all grains and dairy, as well as simple carbohydrates and sugar, all of which are inflammatory. He then details which foods and spices people should eat more, stating, "simple adjustments in our food choices can provide profound healing."[4]

Furthermore, he provides a "Brain Food Pyramid" chart as a reminder. For example, studies have found beneficial cognitive effects from turmeric/curcumin, saffron, and organic matcha green tea. Since an abnormal, unhealthy microbiome has been found to be connected to autoimmune disease, his chapter on improving your microbiome is significant for our purposes. It is becoming increasingly accepted that the microbiome plays a significant role in autoimmunity, as mentioned previously in this book.[9] Bacteria in the gastrointestinal tract have even been tied to decreased cognition (neurodegeneration).[10]

Regarding exercise, Bredesen states, "being active is the single most important strategy you can employ to prevent and remediate cognitive decline," and it has the most scientific evidence to date, but, as he points out, exercise alone is rarely enough. This is consistent with studies quoted earlier in this book. Importantly, physical exercise stimulates the production and release of BDNF (brain-derived neurotrophic factor), encouraging neuronal growth and bringing oxygen to the brain.[11] It will have a greater effect on your memory the earlier in life you start to exercise.[12]

Moreover, Bredesen's book also addresses various options to lower stress and many ways to challenge your brain, including the BrainHQ computer exercises mentioned earlier. For me, the best part of Bredesen's book was the epilogue, in which he states he firmly believes complex chronic illnesses, such as lupus, ulcerative colitis, rheumatoid arthritis, Alzheimer's, and many more, will end in this century. What hope that provides for our suffering, as well as for our children, and grandchildren!

O'Bryan's Approach

I would also recommend the book *You Can Fix Your Brain* by Tom O'Bryan.[13] I think this book is easier to follow and understand than Bredesen's book, although O'Bryan refers to Bredesen's work, and both books contain excellent information. O'Bryan starts the book by saying there are four main components to approach health: structure (physical therapy, chiropractic care, etc.); biochemistry (nutrition, air, toxins, immune system, etc.); mindset (emotional/spiritual aspects); electromagnetics (exposure to toxic EMFs). He states, "a great brain begins in your gut," discussing how intestinal permeability (leaky gut) can lead to systemic inflammation and a leaky/inflamed brain.

Indeed, autoimmune disease has been tied to gut problems (intestinal permeability and dysbiosis) in many research studies over the last decade. In many people, a leaky gut stems from the consumption of gluten, although dairy and sugar can also cause significant problems. In fact, O'Bryan states, "if you want to think more clearly, avoid sugar" because "eating processed sugar increases systemic inflammation." In addition to a leaky gut, patients should also be tested for gut dysbiosis (imbalance of microbiome with too many bad bacteria), which can also negatively affect the brain.

He goes on to discuss the importance of detoxification as a component of health, encouraging patients to eat lots of cruciferous vegetables (broccoli, cauliflower, bok choi, etc.) to help their bodies get rid of toxic chemicals they are exposed to every day (see ewg.org for specific information on products). Patients should educate themselves about such chemicals and remove all harmful substances from their home/work environments. Green tea and specific supplements can also help methylation (vital chemical and biological processes).

O'Bryan also introduces some specific tests your doctor can do for biomarkers to detect a leaky brain or what he calls B4 (breach of the blood-brain barrier). Interestingly, he says if you're sensitive to gluten, you may also have elevated antibodies to your cerebellum. This struck a chord with me as I've had gluten sensitivity for years, and then, lately, my balance has deteriorated, and I'm showing signs of cerebellar ataxia (difficulty coordinating balance & gait). For a thorough deep-dive into leaky gut, inflammation, and gluten issues, see published studies by Dr. Alessio Fasano.[14]

These issues are vital to the health and well-being of autoimmune patients. O'Bryan addresses structure later in the book, reminding us that if there is a problem in the alignment with our spine or joints, the misalignment can affect nerve function, which can then alter organ health. He even says chronic low back pain can lead to brain inflammation. He offers tips on posture and how to sleep. Similarly, he addresses mindset by pointing out how the microbiome influences stress and how important it is to get out of a "sympathetic-dominant state" to work toward a parasympathetic state, where we have a calm mind and good digestion. He gives specific steps to help patients achieve the parasympathetic state. Besides avoiding gluten, dairy, and sugar, O'Bryan gives specific advice on what to eat: dark-colored fruits, omega-3 foods, cinnamon, and parsley, for example. His last main chapter discusses research related to EMF (electromagnetic field) pollution and the negative effects of cell towers, phones, and Bluetooth. Emerging evidence ties EMF exposure to cognitive decline[15] by influencing calcium signaling and increasing inflammation, so you should consider reducing your exposure.

The best thing about his book is that he gives you actionable weekly steps on improving your brain health, which keeps everything from becoming too overwhelming. I highly recommend this book for patients struggling with cognitive issues. It would be even better to enlist a friend or family member to read these types of books with you, discuss them, and decide on specific actions or changes you can make. That way, you will have both support and accountability.

Hyman's Approach

For additional diet information, I highly recommend Dr. Mark Hyman's books *Food: What the Heck Should I Eat?*[16] and *The Pegan Diet.*[17] These books detail what makes up a great healthy diet and tell you what to eat and avoid. Dr. Hyman talks about the 75% rule: you should fill 75% of your plate with non-starchy vegetables. He also tells you the best ways to prepare vegetables and meat. He advocates healthy smoothies to add more veggies to your diet. I often drink a smoothie of veggies, seeds, fruits, protein powder, and superfoods (based on his recipe) for breakfast or a snack. An older book of his, *The Ultramind Solution* (2010),[18] is also an excellent resource, connecting cognitive fog, anxiety, and depression to nutritional factors.

Nutrients

For ideas on supplements and other dietary factors to consider, please refer back to chapter 11, as research has indicated cognitive benefit from B vitamins, DHA/EPA, and flavanol (flavonoids and catechins and procyanidins), as found in grapes, cocoa, and tea.[1] In addition, as O'Bryan indicated (above), you might want to consider polyphenon E (main component EGCG antioxidant) from green tea, as research indicates a neuroprotective effect.[19] Two to four cups of tea (green, black, white, oolong, or dark) a day has been associated with a reduced risk for cognitive decline and a supportive immune system.[20]

Research shows that adding flavones and anthocyanins to your diet will help protect your cognitive abilities.[21] Also, consider adding senolytic agents (to fight cellular senescence or aging of your cells), like quercetin, fisetin, piperlongumine, and curcumin.[22]

To put it simply, consciously choose colorful vegetables and fruits daily, such as orange, yellow, blue, black, and red. Also, don't forget to look into research on the MIND (Mediterranean-DASH Intervention for Neurodegenerative Delay) diet, as research has shown greater adherence to this diet to be associated specifically with better verbal memory, as well as better global cognitive function and larger brain volume.[23] The MIND diet has 15 components, advocating ten healthy food groups and discouraging five unhealthy groups (sweets, fried/fast food, cheese, margarine/butter, and red meat). Older adults who followed the MIND diet were tracked for years and were found to develop "cognitive resilience," resisting cognitive decline due to their diet.[24]

You may also explore information on the components of the Mediterranean diet and DASH (Dietary Approach to Stop Hypertension) diets, since they have both been associated with decreased cognitive decline in other studies.[25,26] In addition, the Autoimmune Protocol Diet (AIP) was designed specifically for people with autoimmune diseases. For more information on the AIP diet, please see https://autoimmunewellness.com/opt-in/.

A recent news article pointed to new research linking an anti-inflammatory diet (with lots of vegetables, fruits, beans/legumes, and tea/coffee) to a decreased risk of dementia.[27] The diet plans mentioned above are anti-inflammatory, so explore them and choose the best one for you.

In contrast, if you eat a highly processed diet (with refined carbohydrates, saturated fats, and low fiber), you are likely causing inflammation and damage to your hippocampus and amygdala, impairing memory and decision-making.[28] Remember that a higher intake of ultra-processed food has been found to increase the risk of lupus by over 50%, so it follows that such a diet would only increase lupus symptoms.[29] A pro-inflammatory diet was found to be associated with an increased risk for dementia in a study of over 1,000 individuals.[30] Don't forget to refer back to chapter 11 of this book on supplementation since adding nutrients like selenium to your diet (found in Brazil nuts) may also positively affect your cognition.[31]

Challenge Your Brain

Let's turn our attention now from diet to brain exercise. Just like diet, brain exercise is also under your control, so you can begin to make changes now to improve your brain health. It is your job to challenge your brain. Your thoughts can change your brain structure. How you exercise your brain is up to you, as long as what you choose requires intense focus and is not too easy for you. To motivate you, let me assure you that this works.

Brain plasticity, the ability of the brain to change, is now accepted as fact. (The brain was once thought to be fixed in adulthood, but now we know new neurons form, as well as new connections, in response to what we are learning.) If you want to understand the concept of neuroplasticity better and read the research to back it up, I highly recommend Norman Doidge's book "The Brain That Changes Itself: Stories of Personal Triumph from the Frontiers of Brain Science" (2007) for an easy-to-read overview. Eric Kandel won the Nobel Prize in 2000 for showing that our neurons alter their structure and strengthen synaptic connections between them as we learn.[11]

Brain HQ, as I've already stated, is one research-backed way to challenge your brain daily. Moreover, learning a new language is another research-backed way to challenge your brain. Learning new dances is also an incredible challenge for your brain, as is learning to play an instrument. Another avenue is embarking on a new, unfamiliar career or intensely reading and studying a difficult subject. When something becomes easy and routine for you, then it's no longer a challenge for your brain, and you must find something new to attempt.

Dr. Michael Merzenich, who developed BrainHQ, says the problem with cognitive impairment is that the brain is noisy due to atrophy of the nucleus basalis and attentional system.[31] We can battle such noise through brain exercise. The nucleus basalis secretes acetylcholine, which helps the brain tune into things and form sharp memories, but people with MCI have little to no measurable acetylcholine in this area. Brain exercise can slow cognitive decline and improve memory, problem-solving, and language functioning.

Remember, massed practice is best, in which you train frequently for short, intense periods rather than long, infrequent periods. One recent study compared the benefits of utilizing BrainHQ to using the app Duolingo (which teaches you a new language).[32] Both were found to increase cognitive abilities, particularly working memory and executive function, but BrainHQ was better at improving processing speed (since it has time components).[32] You may also want to try the app Lumosity, which lets you play three free brain games daily (https://www.lumosity.com/en/).

In addition, don't underestimate the power of daily reading. A recent study found that daily reading improved working memory, episodic memory, and sentence processing in adults over 60.[33] Of course, you certainly don't have to be over 60 to experience the benefits. That study examined reading of fiction, but I recommend that you also spend time each day in nonfiction, studying something you don't already know. The best thing you can do is make a commitment to set aside at least 30 minutes per day, five days per week, to challenge your brain.

Before beginning your daily brain training, you might want to know where your cognitive level is now. You could pursue a full cognitive evaluation through a speech-language pathologist or neuropsychologist. A cognitive evaluation is an excellent option, but be aware it may be costly and not covered by insurance.

Alternatively, you could request a cognitive screening at your rheumatologist's office. If that is unavailable, you can explore online options, such as BrainTest, which is based on the evidence-based SAGE (Self-Administered Gerocognitive Examination).[34]

Moreover, it can be completed online: https://braintest.com/subscription-page/ .

Another option is the XpressO by MoCA app (https://portal-us.mocacognition.com/), which allows you to screen your cognitive function at home. Otherwise, an application like BrainHQ will track your progress and continually challenge you when you utilize the paid version. Other brain training options include

Elevate (https://elevateapp.com/research) and Lumosity (https://www.lumosity.com/en/), although I couldn't find as much research to support them as I could for BrainHQ. (https://www.brainhq.com/?v4=true&fr=y)

Physical Exercise

I know you want to, but you can't forget physical exercise. Numerous studies link better cognition to exercise—do a quick Google Scholar or PubMed search. The exercise-cognition connection is now so widely accepted that the World Health Organization and the American Academy of Neurology both recommend exercise to decrease cognitive decline.[35] "Exercise-induced structural changes result in increased efficiency of neural activation and communication."[36] In this review article, aerobic exercise was consistently found to induce neural changes and increase neuroplasticity (positive brain changes). They only investigated aerobic exercise in healthy adults but acknowledged other studies have found similar results in patients with illness. They also point to a group of studies showing benefits with strength training, although that type of exercise induces a different kind of neural change.[36] BDNF (brain-derived neurotrophic factor), which is important for neuronal growth and vitality, was found to increase during both light and intense aerobic exercise, although the increase was greater with high-intensity exercise.[37]

Adults who combine moderate physical activity (cardio) and weight training in their exercise routines will gain additional benefits, although cardiovascular exercise afforded the most benefits in one study about reducing all-cause and cardiovascular disease mortality.[38] Patients who add in an exercise involving limb coordination, such as learning a new dance[39] or practicing Tai Chi[40] will benefit their brains more. A combination of exercise types is beneficial for cognition. However, you must start where you are and do what you can, whether it's gentle stretching or walking. Walking only three times a week has been shown to increase memory,[12] so that's a great way to start. For additional research on exercise and cognition, refer back to chapter 12.

Stress Management

Stress management has been mentioned but deserves more attention. Managing stress is another important aspect of your health and one that is largely under your control. Stress reduction can take many forms. "A growing body of research shows that meditation has positive effects on cognition in younger and middle-aged adults."[41] Moreover, it is thought to offset age-related cognitive decline.

Meditation can take many forms, of course, with various deep breathing techniques or prayer. It is also beneficial when tied to movement, as in yoga, Tai Chi, and Qigong. I highly recommend you find a good form of daily stress management that works for you. You can find resources in books, articles, and even YouTube.

In addition, there are a number of new biohacking devices (technology designed to benefit your health) on the market, which claim to help you manage your stress and learn how to stay more in your parasympathetic nervous system (the calm, rest & digest system) and less in your sympathetic nervous system (the fight, flight, or freeze system). Some also help improve your heart rate variability, which is an important health factor. I can't yet recommend all these devices as backed by sufficient research, but they may be worth exploring for yourself: explore Braintap

(braintap.com), Oura ring (ouraring.com), Heartmath Inner Balance (Heartmath.com), Apollo Wearable (apolloneuro.com) etc. Heartmath appears to have the most research behind it.

Finally, over 20 studies have emerged showing the benefits of earthing or grounding on stress management and decreasing pain for some people.[42] This method is as simple as putting your bare feet on the grass or sand for a few minutes each day, but results can also be achieved with indoor grounding mats and sheets. To review this topic and explain how the earth's powerful energy can benefit health, read the book by Ober, Sinatra, & Zucker.[43] You can also watch the earthing movie (https://www.earthingmovie.com/) and decide whether or not to try it. Related to this, being near "green space" has recently been associated with higher cognitive and psychomotor speed/attention scores in over 13,000 women.[44] So, being in nature is very good for your brain health.

Your Home Program

To summarize, a good home program would require you to slowly incorporate various lifestyle changes into your daily routine. You might start with your diet, for example. If your doctor agrees, you might try eliminating gluten first and then dairy, and consider whether nightshades (a certain group of vegetables) should be on your list of "never" foods. Dramatically cut or completely eliminate your sugar intake, as sugar causes inflammation.

In addition to cutting things out, concentrate on adding more good foods into your diet, such as cruciferous vegetables, dark leafy greens, and polyphenols (berries). The more sources of polyphenols (which include phenolic acids, flavonoids, stilbenes, and lignans), the better neuroprotection you will have for your brain, so investigate multiple sources found in spices/herbs, fruit, vegetables, nuts/seeds, and certain beverages.[45]

A recent randomized study, for example, found blueberries to improve cognition and help prevent neurodegeneration in middle-aged adults.[46] An additional study found multiple benefits of blueberries for the brain.[47] Walnuts, too, are associated with better cognition and less decline.[48] Furthermore, a systematic review of studies on cinnamon found memory and learning benefits to cinnamon.[49] Find an anti-inflammatory diet that is healthy and works for you, but eat as much organic, whole food as possible–not boxed or processed food. If you see no progress, consider seeing a functional medicine doctor and a dietician for food sensitivity testing and nutritional assessment. If you want to incorporate even more nutrition, you might try one of the supplements listed previously as beneficial for your brain. If you choose to do this, it is recommended that you only add one at a time (with your doctor's approval).

Make challenging your brain a **daily** priority. If something is easy for you to do and think about, it won't work. My first recommendation would be to download the app Brain HQ, but you can also add Duolingo (a language app), crossword puzzles, and other brain games. You can also read a chapter of a difficult-to-understand book daily and/or take up a musical instrument. Don't forget to also add exercise 5-6 days per week. If exercise is difficult, start with Yin Yoga (very slow stretching) and walking when you're fatigued. For strength, you can work up to more advanced yoga, Pilates, water aerobics, and light weights or bands. Even if you feel completely awful, try to stretch some daily. Also, make sure you practice daily prayer and/or meditation. Meditation is important for your emotional, spiritual, and physical health. Finally, socialization is great for your brain, so make socializing a priority every week, even if it has to be by FaceTime or Zoom for health reasons. If you want to go further, start eliminating chemicals from your home. That may seem like many changes, but as Dr. O'Bryan recommends, commit to making one small change at a time.

Find Balance in Your Body

Throughout all my research, the theme that kept emerging was balance. I hadn't expected this, but I began to realize becoming healthy is all about balance on a broad scale. At the root of all autoimmune diseases is a lack of balance. We want our immune system to work enough to fight off unhealthy viruses and bacteria, but we don't want our immune system to overwork and attack our healthy tissues. "Immune balancing is more important than immune boosting" for autoimmune disease.[13] Similarly, although we're often told chronic stress is bad for us and we should avoid it, scientists are now discovering a certain amount of good stress is beneficial (if interested, look up "hormesis"), as in a brief exposure to an extreme temperature, or a controlled period of fasting.

Inflammation also requires balance in the body, as the acute inflammatory process is very important to healing our bodies after an injury. Still, persistent, chronic inflammation leads to disease. Our nervous systems crave balance, as activation of the sympathetic nervous system (and then release of stress hormones like cortisol) is an important survival mechanism in an emergency, but chronic, repetitive overactivation of the sympathetic nervous system prevents the person from being able to slip back into the normal, parasympathetic rest and digest state (the state at which our body works optimally).

Even at the cellular level, balance is important; as studies show that cytokines are critically important to synaptic function and memory, but too many pro-inflammatory cytokines can damage the brain. Similarly, certain antibodies, such as anti-NMDAR and anti-P antibodies, must have concentrations delicately balanced so they are not too low nor too high to achieve optimal cognition, as related to the hippocampus and amygdala.[50]

In our microbiome, we must have a balance of various kinds of bacterial species so that dysbiosis and leaky gut are not triggered (which is one major theory for the cause of autoimmune disease). Our bodies need balance, or homeostasis (a stable internal environment), to work properly. Even on a conscious level, however, balance is important, as, when thinking about our own lifestyle choices, it is clear we should exercise, but if we overexert our bodies, it can lead to an increase in cortisol and additional systemic inflammation. It has also become apparent that, as autoimmune patients, we must listen to our bodies regarding fatigue or lack of energy. If we push ourselves beyond what we are able to do on a particular day, our bodies suffer for a prolonged period.

We cannot overestimate the importance of optimal sleep (not too much or too little) for both body and brain health. Further, working and challenging our brains is necessary. Still, we must balance our challenging work with periods of rest, whether that includes meditation, stretching, napping, prayer, or simply sitting outside for a nature break. Even with eating, there must be a balance in the nutritious foods we choose. We must have a **variety** of types (eat the rainbow) for optimal health, eating various kinds of fruits and vegetables while getting enough protein and healthy fats, so that we're balancing all the components of nutrition. Organic fruit is nutritious, for example, but you can actually eat too much fruit, which then causes your blood sugar to get out of balance. It seems that balance is the key to optimal health.

Even conventional and functional medicine requires balance in my life, as I still lean on a couple of medications but have eliminated the need for many of them by adjusting my diet and nutrient levels. My point is that, as a patient with chronic disease, you need to intentionally think about balance and work toward getting back the homeostasis in your body. This process involves balancing your mind/emotions and your whole life. You CAN make a difference in your own health. You make many choices every day, so choose well for your health. Don't be a passive patient; do not do whatever your doctors tell you without questioning. Seek out information, read articles and books, and ask lots of questions. Take control as much as possible, and never, never, never give up!

✶ ✶ ✶

Special Considerations
Stress plays a powerful and often underestimated role in the lives of those with autoimmune conditions. Unlike their healthy peers, autoimmune patients must consider how emotional strain, work

demands, and uncertainty about the future uniquely affect their health and cognitive function. Scientific evidence continues to show that chronic stress not only worsens autoimmune symptoms but can also physically alter the brain, with elevated cortisol levels contributing to cognitive decline.

This section takes a closer look at the effects of stress, the psychological impact of cognitive diagnoses, challenges at work, and the importance of thoughtful future planning. By addressing these issues with sensitivity and offering practical strategies, we can help patients navigate these complex concerns while maintaining control and hope.

Issues with Stress
Stress, work, and future planning are special considerations in the lives of autoimmune patients, which may differ from those of their healthy peers. Stress management was briefly mentioned previously as one of the four pillars of cognitive health in Masley's book *The Better Brain Solution: How to Start Now at Any Age and Prevent Insulin Resistance of the Brain, Sharpen Cognitive Function, and Avoid Memory Loss.*[6] The impact of stress should not be overlooked, as most patients with autoimmune disease have witnessed the negative effects of stress on worsening disease. In addition, though, be aware there is an established connection between increased cortisol (stress hormone) and a negative physiological impact on brain structure. Calm brains are happier, more effective brains. Both oxytocin (the cuddling hormone) and endorphins (released with exercise) can help calm the brain and help you become mentally sharper.[6] Through the connection between exercise and stress control, we again see the importance of combining various therapies/treatments or life choices to improve cognition.

An additional concern that has been raised in the general literature about cognitive changes in aging adults is the potential stress of a diagnosis of MCI on the patient. Could telling a patient that he/she has MCI or cognitive dysfunction cause more stress in the patient's life? Would worrying about the diagnosis potentially increase symptoms of the disease in lupus patients? A high percentage of lupus patients have already been found to have increased levels of anxiety and depression, so these factors must be considered.

Along these lines, 46% of people living with undiagnosed dementia said that fear of diagnosis and stigma were two reasons they didn't seek a diagnosis.[51] Perhaps such questions need to be explored in research with lupus patients, likely with patient questionnaires. I would hypothesize, though, that professional clinicians (whether speech-language pathologists, neuropsychologists, or neurologists) have been trained to deliver the news of this type of diagnosis to each patient in a gentle, compassionate, and educational way, while simultaneously offering the patient hope through treatment options and support along the way.

Recently, a study was done in which researchers provided patients (and providers) with carefully prepared, color-coded reports of the patient's cognitive scores and questioned them on the format.[52] Overall results were favorable, but patients tended to prefer the function of the report as a communication tool in order to open up discussions with their providers about cognitive struggles. In contrast, providers thought they would utilize the information to inform themselves on how best to communicate with their patients based on the patient's level of cognition. Both groups stressed the importance of using "non-stigmatizing language" when conveying cognitive results to patients. For example, they did not appreciate words like "below average" or "impaired" but preferred terms like "fair" or suggested just color-coding the results instead of using words or listing actual numerical scores[52].

Further research like this would be beneficial in discovering the best ways to inform patients of the extent of their cognitive difficulties without further stressing them. If help is available immediately, such results can be easier to process positively. Suppose patients are immediately aware that there are treatments to help them, and they can be referred to a clinician for help (such as an SLP or neuropsychologist). In that case, they are more likely to accept a diagnosis of cognitive impairment with less anxiety.

Work Concerns
We know that cognitive difficulties impact work, but the impact can vary depending on the cognitive level required of a specific occupation. Vocational counselors may be helpful if a patient is considering modifications or changes in their employment. For patients who are just beginning to struggle with work, please be aware of your rights under the Americans with Disabilities Act (ADA) (https://adata.org/learn-about-ada). In addition, consider talking to an advocate from the Job Accommodation Network (JAN) (https://askjan.org) if you need guidance before talking with your employer about needed modifications.

Part of the reason patients may be reticent to admit to cognitive difficulties is the fear of losing their jobs and, thus, their livelihood. How will they make a living if someone decides they can no longer do their job? A quote from a physician discussing cognitive dysfunction in his patients states, "A lot of our patients are, surprisingly, professionals who are still doing things like teachers or pharmacists, so some of their daily work is very important. If we found something like this (cognitive impairment), it might be something we would have to intervene on."[52] First of all, I don't know why it would be "surprising" that lupus patients are professionals and, second, every person's daily work is important, no matter their profession. What about the fear that your job would be taken away if you admit to (or are diagnosed with) mild cognitive impairment? As a patient, I find this prospect absolutely terrifying! It highlights the importance of education for providers and patients on the nature of cognitive impairment. In addition, it screams the need for evidence-based remediation and accommodation techniques for lupus patients with cognitive impairment.

Future Planning

Another difficult area of conversation arises in the discussion of future planning with patients who have cognitive dysfunction. Early future planning is recommended by the AAN (American Academy of Neurology) since we don't know whether cognitive deficits will progress to dementia in patients (or resolve or stay consistent).[35] If cognitive dysfunction does progress to dementia, however, difficult subjects, such as a patient's end-of-life wishes, advanced directives, estate planning, etc., should be discussed with each individual early in the process.[52] Worsening dementia may not end up being a factor with autoimmune patients, but these issues should be considered, just in case.

In addressing autoimmune diseases, particularly those affecting cognitive function like lupus, the importance of a holistic approach cannot be overstated. The convergence of multiple treatment strategies—including lifestyle modifications, dietary changes, stress management, and cognitive training—has shown promising results in improving brain health and slowing cognitive decline. Research now supports that a combination of interventions, such as diet, exercise, cognitive training, and emotional wellness, is more effective than a one-size-fits-all solution. Dr. Bredesen's precision medicine approach and Dr. O'Bryan's focus on gut health and detoxification are key examples of how comprehensive, multi-faceted strategies are vital for addressing the complexities of cognitive health.

Moreover, balance remains a crucial theme in maintaining overall health. A body in balance can effectively manage the immune system, inflammation, and stress, while ensuring that the brain remains sharp and functional. Addressing stress—both physical and emotional—becomes increasingly important in preserving cognition, as chronic stress exacerbates autoimmune symptoms and accelerates cognitive decline. As autoimmune patients face unique challenges, they must embrace the power of balance and adopt an informed, proactive approach to their health.

Finally, practical strategies for managing cognitive changes—such as a focus on brain exercises, regular social interaction, and mindful self-care—are essential in maintaining mental clarity and emotional stability. By adopting these strategies and working closely with healthcare providers, autoimmune patients can improve their quality of life, reduce cognitive impairment, and foster a sense of hope for the future.

Part IV:

Exploring Rehabilitation

Chapter 16

What are the Types of Cognitive Rehabilitation?

Now that the various treatment options have been discussed, we will focus on specific types of cognitive treatment. Confusion remains regarding the terminology: cognitive stimulation, training, and rehabilitation.[1] Cognitive stimulation is often used when talking about general thinking activities or things that are good for the brain in general (like reading, crossword puzzles, etc.). Cognitive training would be like a computer program or app that focuses on a specific cognitive domain, such as visual memory. Cognitive rehabilitation would be a more specific and individualized approach that targets goals established by a patient and his/her family with the guidance of a professional speech-language pathologist or neuropsychologist.[2] Cognitive rehabilitation may be beneficial in improving the quality of life of lupus patients.[3] When all of these approaches are combined into one program, that would be the most advantageous scenario.

Importance of Individualized Programs

Designing an individualized program is an important consideration in determining the focus of treatment goals. Perhaps the main goal for a particular patient is to restore cognitive function to what it used to be or to improve the level of a particular function. Alternatively, the goal may be to help a patient compensate for deficits that cannot be improved or restored, with things like supportive technology or aides discussed in a previous chapter.

The ultimate goal for a patient is not the improvement of a standardized score on a cognitive subtest, but the ability to function more independently in daily life. According to patients and families, research has shown that the most important outcomes are quality of life and self-efficacy.[4]

Moreover, it is recommended that clinicians focus **both** on restorative cognitive training (with structured, repetitive practice of specific tasks) to address mild cognitive deficits **and** compensatory cognitive training to address mild difficulties with daily functioning and teaching relaxation/stress management strategies to address commonly associated neuropsychiatric issues.[5]

Combining EBP Techniques with Lifestyle Changes

Evidence-based practice (EBP) guides clinical decisions based on evidence from research and thus improves the quality of treatment given to patients.[6] For an explanation of EBP levels of research and how to categorize types of research, please see the Oxford Center for Evidence-Based Medicine guidelines.[7] Future research needs to tell us not only what the best types of cognitive intervention are but what **combination** of approaches might work best, in addition to the dosing of treatment (how long and how often) and whether individual or group treatment works better.[1]

For example, a realistic scenario might include a lupus patient with cognitive issues seeing an SLP for individualized sessions once a week while completing recommended computerized cognitive exercises five days a week and utilizing compensatory technology daily. That patient's program might also include regular exercise five days per week, a healthy diet, and a couple of supplements chosen by his/her rheumatologist or functional medicine doctor, in addition to any necessary medications. Research shows that patients want such a holistic approach to treatment.[8]

A review of the literature on cognitive rehabilitation for Alzheimer's disease shows cognitive rehabilitation has greater benefits when combined with other types of intervention.[6] One study, for example, found that adults with MCI made more gains with paper crossword puzzles vs. computerized brain games, but upon further analysis, they realized that adults with more severe impairment were more comfortable and familiar with paper puzzles, so they did those more.[9] In searching for evidence-based treatments, researchers must not forget why they are doing this work: to help patients achieve the individualized outcomes they need. Patients must be able to live independently and successfully as much as possible.

Useful Field of View Approach
Now, let's turn our attention to very specific types of cognitive rehabilitation. For example, a systematic review of "Useful Field of View" training in older adults found that this type of cognitive training improved processing speed and attention, and gains remained after ten years.[10] The useful field of view is the visual area where we can gain information without turning our heads or moving our eyes, and it usually decreases in accuracy with age. Such a skill relates directly to daily life activities, such as driving. Thus, training in this area can positively affect a patient's daily life.

Supporting Memory
Researchers have demonstrated certain cognitive rehabilitation interventions initially designed for one population (such as TBI) may be successfully adapted to other patient populations.[11] Again, I mention this since we must adapt specific interventions from other populations and utilize them with lupus patients until we have enough specific EBP (evidence-based practice) literature on cognitive dysfunction treatment options in lupus. To give you an idea of another type of treatment utilized in a therapist-led cognitive intervention rehabilitation program, we will turn to memory.

Memory is a common deficit in both lupus patients with cognitive deficits and patients in the general population with cognitive deficits and is, therefore, often discussed in the literature.[6] Memory is needed daily to learn and recall what was previously learned. Providing hope, a systematic review found cognitive rehabilitation, focused on both memory strategies and supports, resulted in significant functional gains in patients' lives, and the authors recommended clinicians "consider a systematic approach to teaching and implementing memory strategies/supports with their clients." [5]An eight-week "psychoeducational group intervention" for lupus patients with cognitive dysfunction found that "metamemory and memory self-efficacy" improved significantly with treatment.[12]

Spaced-Retrieval and Errorless Learning Approaches

Two other specific approaches are Spaced-Retrieval Training and Errorless Learning. Spaced-retrieval training pairs new associations through repeated trials, while errorless learning attempts to minimize errors while breaking down complex processes into short steps.[13] In Spaced-Retrieval, the patient carries out a procedure or recalls information while increasing the time intervals. This technique also utilizes behavioral shaping (breaking a task into small steps) and cueing.[14] The Errorless Learning technique attempts to have patients avoid making mistakes so their incorrect responses are not reinforced.[14] These two types of cognitive training techniques for new learning and memory have been successful with patients with mild dementia.

For in-depth instructions and research on Spaced Retrieval, see "Spaced Retrieval: Step by Step" by Benigas, Brush, and Elliott.[15] They break down the procedure of working with a patient to address very specific memory tasks. Initially, the clinician must identify a particular, individualized need (i.e., steps for how to charge your cell phone or steps for how to stand safely). The clinician then prompts the patient to repeat the instructions repeatedly at longer and longer intervals. The patients practice performing the action(s) while verbalizing the steps. Visual cues, as well as other cues, are added as needed. The key to the success of the Spaced-Retrieval approach appears to be lots of repetition throughout the day, focusing on one piece of information to remember.

Sohlberg & Mateer's Approach
Sohlberg and Mateer have effective techniques for various cognitive domains (not just memory), and their training protocol involves acquisition, application, and adaptation.[16] Their groundbreaking book on cognitive rehabilitation takes a hierarchical, organized, process-specific approach to treatment. Remediation is data-driven, and generalization (outside of therapy sessions) is an important priority.[16] They focus on repetitive therapy tasks to help restore lost cognition but also advocate teaching compensatory strategies.[17] They outline specific approaches to remediate the following cognitive areas: orientation, attention, memory, visual processing, language impairment, executive function, and reasoning/problem-solving.[17] Since I have utilized Sohlberg & Mateer's approach with my cognitive patients, I highly recommend their Attention Process Training (APT) method. I have found this type of treatment to lay an important foundation for success in memory training with my previous patients.

Speed of Processing Training
Speed of processing training[18] was found to decrease the chances of dementia development in typical older adults by 29%. This program focused on utilizing computerized visual-perceptual exercises to increase how fast the individuals could process information. The amount of information and the complexity of the information both gradually increased. Interestingly, researchers found their computerized memory and reasoning training did not have the same positive effects.[18]

Word Retrieval Techniques
Suppose anomic (difficulty recalling names of objects/actions/concepts) deficits are present. In that case, many

word-retrieval techniques, familiar to speech-language pathologists, have been utilized effectively with patients with aphasia, TBI, dementia, etc. Some of these include semantic feature analysis, massed practice using generative naming, and repetitive naming with semantic descriptions.[14] These three treatments assume the breakdown results from semantic memory deficits. Other commonly available techniques to address word retrieval include orthographic cueing (cueing with written words), phonemic cueing (providing sounds to cue the word), and visual cueing with pictures.

General Language Stimulation
General language stimulation interventions are another option. One example of general language stimulation is reminiscence therapy, in which themes or topics from the past are utilized in treatment. This approach was developed after researchers found retrograde memory (before cognitive decline) is often better preserved than anterograde memory (creating memories with new information).

While the above treatment techniques are far from comprehensive, they give you an idea of the types of evidence-based cognitive intervention techniques that are currently available. Unlike the lifestyle factors addressed in chapter 15, a clinician, rather than the patient, would choose which treatment technique might be appropriate for a particular patient. Technique selection is based on the patient's goals, as well as the patient's testing results, which highlight specific cognitive deficit areas. The next section will be addressed to speech-language pathologists, neuropsychologists, or other clinicians working directly with autoimmune patients experiencing cognitive impairment.

A Note to SLPs and Other Clinicians
If you are a speech-language pathologist (SLP), neuropsychologist, or other clinician addressing cognition, you may have never previously considered treating lupus or other autoimmune

patients. However, if you work in a medical setting, you have most likely worked with other types of patients with cognitive deficits (such as patients with dementia, patients with CVA, patients with TBI, etc.). Be aware various studies show people with chronic illness tend to have more cognitive impairments.[19] Remember, "SLPs can play a key role in dementia management, particularly with respect to identifying and providing suitable behavioral or nonpharmacological intervention."[14] ASHA's (American Speech-Language-Hearing Association) stance on cognitive treatment for dementia is positive.

Dementia is a rising population, as over six million Americans have Alzheimer's disease, and those numbers are expected to rise to over 13 million by 2050.[20] Also, 75% of patients with dementia globally are undiagnosed.[21] Therefore, one reason for this is a lack of trained clinicians for diagnosis and assessment. It is also appropriate for SLPs to administer cognitive treatment to patients with milder conditions, like MCI. Thus, it would be appropriate for SLPs to treat autoimmune patients with cognitive difficulty. As mentioned previously in this book, there's a paucity of research on specific evidence-based cognitive rehabilitation techniques for patients with SLE. Therefore, clinicians must borrow from evidence-based techniques in other populations. Some of these specific techniques were mentioned above, although the list was by no means comprehensive.

Scope of Practice
You should not hesitate to take on autoimmune patients in your caseload. As a study states, "CI (cognitive impairment) is likely more widespread among patients with SLE than previously expected."[22] If you are in a hospital setting, be aware that, for lupus patients, "cognitive functioning is not well described in most medical charts and often not discussed in the clinical setting."[23] You can be an advocate for change, keeping your eyes open for autoimmune patients with possible deficits. Also, remember that "the opportunity and charge exist for speech-language pathologists

to assert their expertise in the behavioral management of all cognitive-linguistic disorders."[24]

For those of you who have never done cognitive treatment, remember that ASHA (American Speech-Language-Hearing Association) is in agreement with SLPs incorporating cognitive treatment into their practice. It is well within our scope of practice, as SLPs and ASHA offer guidance on coding and billing for services.

For more details visit:

https://www.asha.org/practice/reimbursement/coding-and-reimbursement-of-cognitive-evaluation-and-treatment-services/

If you're interested, make sure you seek appropriate training and do your own research to find the best evidence-based techniques.

Standardized Testing

As with any patient, you would begin with a lupus (or other autoimmune) patient by obtaining a complete patient history, listening carefully to what the patient reports (looking toward patient-reported outcomes), and administering a combination of objective (standardized) and subjective assessments.[25] You can administer standardized cognitive assessments with which you are familiar, but you might also want to consider starting with those used previously in this population, such as the MoCA or the CANTAB.

The CANTAB has been used in various settings and has been found to be appropriate for SLE.[26] The MoCA, described previously in the screening/testing chapter (chapter eight), has been found to be sensitive to deficits in SLE patients and seems to have the best correspondence thus far to the original ACR comprehensive battery.[27] Also, remote cognitive assessment is now an option through the MoCA.[28] In addition, note the optimal screening cut-off threshold for the MoCA in SLE patients is <28.[29] Utilizing the MoCA, researchers have found that 68% of SLE patients have cognitive dysfunction, mainly in the areas of verbal repetition, fluency, abstract generalization, and delayed recall.[27] These results made me wonder if a verbal repetition task might be a good initial screening tool, although research has not yet verified this tool. It is recommended that the MoCA be utilized to monitor cognitive deficits in populations with chronic disease.[30] Also, be aware one study found lupus patients with cognitive dysfunction to have more variability across tests.[31] For more specific assessment information, please see the previous section (chapter eight) on assessment and screening for cognitive dysfunction in SLE.

Goals and Cognitive Targets
Once you identify the patient's weaker areas of cognition, you will know what objectives to write and where to start working. These areas may include memory, executive function, attention, language, and visuospatial skills.[32] According to one study, working memory and language are the domains most affected by lupus, followed by attention and psychomotor speed.[33] Another study found cognitive impairments in lupus patients in the areas of verbal memory, verbal fluency, attention, and executive functions.[34] All lupus patients in one particular study were found to have some type of cognitive impairment, with the most common being the inability to effectively shift attention from one task to another.[35]

In addition, sustained attention is often affected in these patients.[26] Most recently, it was found that 47% of lupus patients have cognitive impairment, with the most severe deficits in the areas of executive function and attention span, followed by delayed recall, processing speed, and new learning difficulties.[36] An SLP can effectively address all of these areas, utilizing methods they are already familiar with in practice. If you need a little direction for cognitive treatment, I highly recommend Sohlberg and Mateer's book for a specific, process-oriented rehabilitation approach.[17] I have found their techniques for attention and memory to be very beneficial. Also, I recommend Joan Green's book *Assistive Technology in Special Education (2018)*, in which Chapter 11 focuses on cognition, attention, and executive functions. For an introduction to various memory treatments, read the article *Evidence-based decisions: Memory intervention for individuals with mild cognitive impairment* by Fleck and Corwin.[5] In addition, you might want to explore the *Cognitive Rehabilitation Manual*[37] or *Optimizing Cognitive Rehabilitation*.[38]

I think language deficits in autoimmune patients have been largely ignored compared to other cognitive domains. In fact, a landmark very familiar to SLPs, Heschl's gyrus (often associated with language processing and comprehension), has been correlated with cognitive issues in SLE patients[39] and may cause memory and "mneumonic impairments." As SLPs, we are very familiar with Heschl's gyrus (Brodmann area 41) and the constellation of symptoms that can arise when this area is affected (as in Wernicke's aphasia). A meta-analysis study revealed that the cognitive domains most affected in SLE patients are complex attention, delayed verbal memory, language, and verbal reasoning.[40] As mentioned, studies have found lexical and semantic memory deficits in lupus patients, although spatial working memory was affected the most. These studies lead me to believe language is an untapped research area in the autoimmune population.[41]

In addition, please note that the gold standard assessment protocol for SLE patients (mentioned earlier in the book) put forth by the ACR does not include an assessment of pragmatic language skills. A recent study found almost half of the SLE patients they studied had pragmatic (social language) deficits. Furthermore, this area was significantly more pronounced than other areas of cognition.[42] Although this is one study and should be verified, SLPs should be aware of possible pragmatic language deficits in their SLE patients. You might consider using something like the Pragmatic Protocol[43] as a guideline for subjective assessment or the Assessment of Pragmatic Abilities and Cognitive Substrates by Arcara & Bambini (2016)[44] for a more objective look at pragmatic language. Also, see Rad (2014) for a review of pragmatic instruments.[45]

Length of Treatment
Numerous treatment techniques are available for cognition and language issues, and a combination of direct rehabilitation treatment, cognitive computer training, and a home program is often the most efficient plan. Although we, as SLPs, would love to help patients as long as possible, there are constraints in the real world, such as a limited number of paid insurance visits. For this reason, patients and clinicians often wonder about the number of sessions required to obtain results. Of course, we are ethically bound never to promise a guaranteed result, as so many individualized factors are at play. One randomized-controlled Alzheimer's prevention study found 12-14 cognitive sessions to be the optimal dose of treatment for most people.[46] Considering this amount of treatment may give us a good starting point in planning treatment, while adjusting to individual needs and utilizing other resources like home programs and computer training to increase efficiency.

Individualized Programs for Cognition
As mentioned earlier, there are currently no agreed-upon lupus cognitive profiles. Each patient has unique cognitive issues, such that each person has an individual "brain fingerprint," which can be verified by multiple sensitive MRI scans.[47] If the patient is seeking help, there is almost always a problem somewhere. Your job is to find the problem (or problem areas) and come up with a great treatment plan. According to a study, "the multifaceted nature of cognitive dysfunction in SLE means any therapeutic interventions should be individually tailored." Fortunately, as SLPs, we are used to designing individualized programs for treatment in all populations.[26] Remember, "now more than ever is the time for SLPs to assertively and proactively proceed on behalf of all adults with mild cognitive impairment."[32]

In addition, be aware of the various symptoms of autoimmune diseases (beyond cognitive dysfunction) and the multiple aspects of treatment required. Educate yourself on multidomain interventions beginning to be explored in conventional medicine[48] and already being utilized in functional medicine.[49] This way, you will be better able to counsel patients on additional resources and lifestyle steps needed (like nutrition, exercise, and meditation) to further improve their cognitive results. Read chapter fifteen for an overview, and refer them to a certified functional medicine health coach to help implement lifestyle changes. Taking control of modifiable lifestyle factors will help patients gain some confidence and a sense of control over their lives, as MCI may have stripped away some of that autonomy. Know also that many, many symptoms come with autoimmune disease, two of the primary ones being fatigue and pain. This will undoubtedly affect progress, so be prepared for inconsistency. With the patient's permission, encourage the patient's family and friends to be involved and supportive for the best possible outcomes.

In conclusion, individualized cognitive rehabilitation programs are essential for patients with cognitive deficits, including those with autoimmune conditions like lupus. These programs should be tailored to the patient's specific needs, whether focusing on restoring cognitive functions, compensating for deficits, or addressing neuropsychiatric issues. Evidence-based practices guide clinicians in making informed decisions that enhance patient outcomes, aiming for greater independence and improved quality of life. Combining cognitive interventions with lifestyle changes, such as regular exercise and a healthy diet, can further support the rehabilitation process. The use of targeted techniques like Spaced-Retrieval, Errorless Learning, and the Useful Field of View training shows promise in improving cognitive functions like memory, attention, and processing speed.

Moreover, clinicians should consider both structured therapies and compensatory strategies to address deficits across various cognitive domains, such as memory, executive function, and language. Given the prevalence of cognitive impairments in autoimmune diseases, speech-language pathologists and other clinicians can play a pivotal role in treating these conditions. By utilizing a comprehensive, evidence-based approach and continuing to adapt interventions based on patient progress, clinicians can significantly enhance independence and quality of life for patients suffering from cognitive impairments associated with autoimmune diseases.

Chapter 17

Do People With Other Autoimmune Diseases Get Cognitive Impairment?

Similar cognitive issues have been more thoroughly studied in other diseases, particularly, for example, in Multiple Sclerosis (MS). Patients with MS frequently have cognitive deficits, but cognitive deficits are being discovered in patients with other autoimmune diseases, as well. The current thought seems to be that there is a higher prevalence of cognitive dysfunction in MS as compared to lupus. It was observed that patients with lupus have less working memory impairment than MS patients diagnosed with cognitive impairment, but they have more impairment than MS patients without cognitive deficits.[1] Lupus patients, however, appear to have a higher prevalence of cognitive dysfunction than patients with Rheumatoid Arthritis (RA).[2] Research has shown the relative risk for cognitive dysfunction in SLE is indeed greater than the risk seen in RA or healthy individuals.[3]

Multiple Sclerosis

As mentioned, research on MS and cognitive impairment has been extensive, much more so than with lupus. MS is an autoimmune disease in which the myelin (insulation of nerve cells) is attacked, causing damage to the brain and spinal cord. It is an inflammatory disease of the central nervous system, causing motor, cognitive, and neuropsychiatric symptoms.[4] Cognition in MS patients was first investigated in the 1980s.[5] It has been found in almost half of patients.[6] In fact, some researchers believe cognitive impairment may be the most devastating symptom of MS.[7] Due to the high prevalence rate of cognitive impairment in MS, regular assessment of cognitive function became routine in the 1990s. Cognitive dysfunction can occur in all stages of the disease.[8] It can even occur very early before other clinical symptoms.[9]

MS patients have been found to have both working memory and processing speed deficits.[10] Information processing speed is the most common cognitive deficit reported.[11] Another study draws parallels between the cognitive deficits of information processing speed, attention, and working memory in both lupus and MS.[12] Verbal fluency and executive functions may also be affected in MS.[5] Interestingly, daily language and verbal skills are usually spared in adult MS patients,[9] at least at first, but this does not hold true for pediatric patients with MS.[13] Aphasia (difficulty with comprehension and language production) is rare in adults.[14] Demyelization, gray matter atrophy, white matter lesions, and inflammatory cytokines are all thought to contribute to cognitive dysfunction in MS,[15] as well as blood-brain barrier disruption and axonal damage.[16] Brain inflammation is thought to cause synaptic dysfunctions, contributing to both motor and cognitive impairments.[17] Cortical lesions may be linked to early cognitive dysfunction.[18] Research continues to investigate possible causes of brain inflammation in MS, as in a recent article linking the neuroinflammation to peptidoglycan (PGN), which is a component of gram-positive and gram-negative bacteria.[19]

Early treatment with disease-modifying drugs (DMARDs) for MS, along with cognitive rehabilitation, have been found to be effective for cognitive impairment in MS. DMARDS found to be effective include Interferon beta -1a; Interferon beta -1b and natalizumab.[4] Studies have investigated the effects of aerobic exercise on cognitive performance in MS with mixed results.[20] One interesting six-week study found improved cognition (and QOL) with vibration training (standing on a vibrating platform) in MS patients.[21] Since mild cognitive deficits are often undetected at regular appointments,[22] cognitive assessments/screeners must be utilized intentionally, much like I am advocating that cognition be screened in lupus patients. Tools commonly utilized for cognitive assessment in MS include BICAMS (Brief International Cognitive Assessment for Multiple Sclerosis), BRB-N (Brief Repeated Battery of Neuropsychological Tests, and MACFIMS (Minimal Assessment of Cognitive Function in Multiple Sclerosis). It has been proposed that beginning (early) cognitive difficulties may be picked up with the CANTAB (Cambridge Neuropsychological Automated Test Battery) by discriminating between normal and impaired cognitive performance in patients with MS.[4]

Cognitive deficits have been identified in all three forms or phenotypes of MS (Relapsing-Remitting MS, Secondary Progressive MS, Primary Progressive MS), but the two progressive forms tend to be more significantly impaired.[16] In fact, in the two latter forms of MS, patients had more difficulty with immediate and delayed verbal recall, attention, and information processing.[23]

Unfortunately, cognitive impairment in MS tends to progress over time.[17] Memory and executive function deficits, in particular, tend to increase.[24] "Pathological processes progressively modify brain networks essential for functional domains," such as cognition, vision, and sensorimotor skills.[17] The number of patients who develop cognitive issues over time (disease duration) increases.[24] What about patients with MS who have normal cognition throughout the disease? Researchers are studying the contribution of cognitive reserve (CR), in which intellectual factors such as educational level and employment may help protect certain patients from disease-related damage to the brain.[17] Alternately, some studies show risk factors for certain cognitive issues may include gender, age, and genetic predisposition.[25]

As in lupus, cognitive impairment in MS has been found to have a significant negative impact on quality of life, employment status, independence, daily living activities, communication, and treatment adherence.[26] Additional similarities have been drawn between cognitive dysfunction in the two diseases. According to some researchers, MS and SLE share a "similar cognitive profile" with deficits in working memory, processing speed, visual/spatial learning, and memory.[27] Although some patients with each disease (MS and SLE) are free of cognitive dysfunction, both diseases are inflammatory in nature and can involve lesions of cerebral white matter.[27] Some studies have found both diseases to have subcortical patterns of cognitive dysfunction, while others have found demyelination in both.[28] Whatever the cause, it is now recognized that cognitive assessments and follow-ups should be "as much of a priority as the evaluation of physical disability" in MS.[26] Why is this not also the case in lupus and other autoimmune diseases?

Rheumatoid Arthritis

Cognitive impairment has been reported more often in lupus and MS, but it is present in other rheumatic diseases.[29] Several studies have also investigated cognitive issues in rheumatoid arthritis (RA). RA is an autoimmune disease in which inflammation occurs in the synovial tissue and causes damage to cartilage and bone.[30] It is an inflammatory condition and involves autoantibody production, causing progressive disability.[31] Like lupus, it is multifactorial, with genetic susceptibility combined with environmental risk factors to cause the disease. Interestingly, "several mucosal surfaces have been implicated as sites of disease initiation in RA, including the periodontium, the lungs, the gut, and the genitourinary system."[32] One good thing about RA treatment is that nine different biological medications have been introduced in the last 20 years, affording patients with RA more treatment options.[33] However, we don't yet know if these drugs help with the symptoms of cognitive impairment.

A recent study found that healthy lifestyle factors, including a healthy diet, regular exercise, not smoking, not drinking alcohol, and having a good body mass index, could significantly lower the risk of developing RA in women.[34] Another study found worse cognitive performance (working memory and executive functions) along with higher autoantibody levels in patients with RA versus healthy controls. In addition, they found increased S100B protein levels in these RA patients.[35] Likewise, researchers found increased S100B levels in SLE patients with cognitive impairment.[36] RA patients have also been found to have difficulty encoding new information, although not as much as SLE patients.[27] Hippocampal inflammation and reduced volume have been connected with cognitive symptoms in RA.[37] Cognitive impairment in RA patients has also been found to affect compliance and success of treatment.[38]

One interesting difference between SLE and RA cognitive function is age. Cognitive impairment is more likely to be found in older RA patients.[39] Even some very young SLE patients, however, are often found to have cognitive impairment. Could this indicate different causes? While a couple of studies found no increased levels of cognitive dysfunction in older individuals with RA, they appear flawed in that they relied solely on patient self-report rather than objective testing for cognitive deficits.[40,41] Prevalence rates for cognitive dysfunction in RA have ranged from 30-50%, and standardized screening tests of verbal learning, recall, and phonemic fluency have been found to identify impairments accurately.[42]

Another study found approximately 1/3 of RA patients have cognitive deficits, particularly visuospatial learning and memory and design fluency.[43] Over 20% of RA patients were found to have executive function deficits,[44] which likely affect daily functioning. RA patients with cognitive deficits have been shown to have increased levels of inflammatory cytokines (like TNF alpha).[39] A pilot study found that anti-TNF pharmacological treatment reduced cognitive impairment in RA patients, which caused them to hypothesize TNF-alpha was responsible for the reduction of cerebral blood flow.[45] In contrast, biologic DMARDS (disease-modifying anti-rheumatic drugs), like the TNF drug etanercept, have been shown to reduce the risk of dementia significantly.[39] One systematic review of the literature confirmed cognitive dysfunction does exist in RA but is less than that of SLE.[3] Surprisingly, several studies have found RA to reduce the risk of Alzheimer's Disease

(AD).[46] This leads to a new area of research. Some researchers think TNFalpha medications, which many patients with RA take, may actually reduce the risk of Alzheimer's disease, and trials to test this hypothesis are currently taking place.[47] We must remember, however, that cognitive dysfunction and AD are related, but not equivalent, terms.

Additional diagnoses for patients should be considered, as well. People with autoimmune diseases often have other medical issues besides their autoimmune disease. For example, one study showed patients with rheumatoid arthritis (RA), who also had cardiovascular risk factors, had an increased risk of cognitive impairment.[48] It is unknown whether those cardiovascular factors contribute directly to cognitive deficits or are entirely separate issues. Many patients (like me) also have polyautoimmunity or multiple types of autoimmune diseases at once.

One study investigated cognitive impairment in lupus, RA, and MS patients all at the same time.[49] These researchers utilized the ANAM (Automated Neuropsychological Assessment Metrics), a computerized subtests battery. They found that MS patients had the most cognitive impairment, while lupus and RA patients had similar profiles.[49] All three groups, of course, performed worse than healthy controls. Other studies have found hypoperfusion in the frontal lobes to be associated with cognitive impairment in both lupus and RA patients.[50]

Sjogren's Syndrome

There is some evidence for cognitive deficits in Sjogren's syndrome, as well. Patients with Sjogren's syndrome may have similar rates of cognitive impairment as patients with lupus.[51] Sjogren's is an autoimmune disease in which inflammation attacks the moisture-producing glands in the body, causing dry eyes, dry mouth, etc. A recent study found patients with Sjogren's syndrome to have increased MD (mean diffusivity) in white fiber tracts of their brains.[52] The peripheral nervous system (PNS), though, is thought to be affected more often than the CNS in Sjogren's.[53] A recent literature review article found "a high prevalence of neuropsychiatric manifestations associated with SLE and pSD (primary Sjogren's Disease)" and called for multicenter studies on treatment.[54]

Psoriatic Arthritis

Cognitive impairment in psoriatic arthritis (PsA) has not been well-researched. Still, one study utilizing the MoCA for assessment found mild cognitive deficits in almost half of patients.[55] Another study found PsA to be associated with dementia.[56] PsA patients suffer from systemic inflammation and are at an increased risk for cardiovascular events.[39, 57] Therefore, it follows that they would also have an increased risk of cognitive impairment.

Additional Autoimmune Diseases and Cognition

Other autoimmune diseases may also be affected by cognitive deficits. Autoimmune disease is very common, affecting up to eight percent of the population in the United States.[58] Inflammatory bowel disease (IBD), an inflammation-based autoimmune disease of the bowel, has been associated with cognitive impairment.[59] However, not much research has yet been conducted to verify the connection. Ankylosing spondylitis (AS), inflammatory arthritis affecting the joints of the spine, has also been found to be associated with cognitive dysfunction compared to healthy controls.[39] In contrast, systemic sclerosis (SSc) rarely has direct effects on cognitive function.[38] (SSc is characterized by diffuse fibrosis and vascular abnormalities in skin, joints, and internal organs.) Future research should determine whether the causes differ in each disease or whether the etiology is the same. Hopefully, research will reveal numerous effective treatment options for all autoimmune patients.

COVID-19, Brain Fog, & Autoimmunity

COVID-19 is the result of a virus and not an autoimmune disease, but since COVID-19 spread across the world several years ago, there's been a lot of awareness of "brain fog" or cognitive impairment associated with long-COVID. Some individuals experience decreased memory and attention after having COVID, and researchers have found neuroinflammation, microglial activation, and mitochondrial dysfunction in those patients.[60] Cognitive impairment can last months to years in post-COVID-19 conditions and often includes slower processing/response speed, up to two standard deviations below normal controls.[61] In studying people who had been hospitalized for COVID-19 with matched healthy controls, those who had had COVID-19 had decreased volume in several brain regions, cognitive impairment, and serum markers of neuronal injury.[62]

Researchers have also begun to link COVID with autoimmune disease. Following SARS-CoV-2 infection, people have an increased risk for autoimmune inflammatory rheumatic diseases.[63] An extensive study analyzed patients in South Korea and found significantly increased risks of various autoimmune and autoinflammatory connective tissue disorders after having COVID, particularly if their infection was severe or they were unvaccinated.[64] Researchers have even found that "high anellovirus load correlated strongly with SLE, RA, and COVID-19."[65] They also discussed that excessive responses to viruses can create a cytokine storm in both autoimmune diseases and COVID. With the association between COVID and autoimmune disease, and the fact that both are associated with brain fog or cognitive impairment, it begs the question of whether the cause of this impairment has a similar mechanism. Hopefully, future studies will provide answers regarding this connection.

Despite the lack of research in this area across most diseases, it is obvious that cognitive dysfunction is a significant issue in autoimmune diseases. Olah states that "most autoimmune rheumatic diseases have been associated with various degrees of cognitive dysfunction."[38] The chronic inflammation found in RA, SLE, and other rheumatological diseases is a significant risk factor for cognitive impairment, as well as cardiovascular mortality and increased arteriosclerotic cardiovascular diseases.[39, 57] Megakaryocyte (large bone marrow cell) expansion has been found in the peripheral blood of RA, SLE, and Sjogren's patients, which may indicate a common pathogenesis.[66] It has been warned that "cognitive dysfunction is a common symptom among patients with systemic autoimmune diseases, most commonly seen in patients with SLE or Sjogren syndrome."[67] Furthermore, inflammation is a risk factor and a possible therapeutic target for cognitive decline in all autoimmune diseases.[68] Research shows patients with any kind of chronic autoimmune rheumatic disease who have sustained disease activity or organ damage should be evaluated for cognitive dysfunction.[38]

Cognitive impairment is a common issue across several autoimmune diseases, with studies revealing its presence in conditions such as Multiple Sclerosis (MS), Rheumatoid Arthritis (RA), Lupus, and others. The extent and nature of cognitive dysfunction, however, vary between diseases. MS patients exhibit significant cognitive deficits, especially in working memory, information processing speed, and executive function, which can be present even in the early stages of the disease. In contrast, Lupus patients show less severe impairment, although they still experience cognitive dysfunction, particularly when compared to healthy individuals or RA patients. While RA is generally associated with milder cognitive issues, a subset of patients does experience problems with memory and executive function.

Research into autoimmune diseases such as Sjogren's syndrome and Psoriatic Arthritis suggests that cognitive deficits are also present in these conditions, although the evidence is still limited. Moreover, the commonality between autoimmune diseases, such as chronic inflammation, highlights a potential shared mechanism underlying cognitive dysfunction. The overlap between autoimmune disease and cognitive impairment, particularly in conditions like COVID-19, points to a broader area for future research. Ultimately, addressing cognitive dysfunction is essential for improving the quality of life for patients with autoimmune diseases, making it a critical focus for ongoing studies.

Conclusion

What Can We Do To Help?

A Call to Action

An encouraging recent development is that a global advisory committee is guiding The Addressing Lupus Pillars for Health Advancement (ALPHA) project. The aim is to establish global consensus and develop strategies to address challenges that limit progress in lupus (including diagnosis, treatment, and care).[1] ALPHA hopes to develop a "global road map of specific recommendations to address identified barriers." This is the first global, comprehensive initiative on lupus, and they stated, "lupus is a major public health challenge that will require comprehensive measures to transform the research and healthcare landscape."[1] What is encouraging to me is that they assert that "every stakeholder has a role, whether researcher, clinician, biopharmaceutical representative, regulatory official, insurer or patient."[1] Patients deserve a much bigger voice since "a patient's perspective is still not accepted as equivalent to the physician's perspective in treatment decisions," as stated by Mosca, who advocates for using patients' reported outcomes (PROs) as treatment targets.[2]

Barriers to Treatment and ALPHA Phase Two

As I write this, phase two of the ALPHA project has just been completed, in which they identified three main barriers to treatment success: drug development, clinical care, and access to care.[3] To address the three barriers (in order), they discussed steroid-sparing, defining the lupus spectrum & biomarker development, and leveraging social media (to address social/cultural disparities in care).[3] If patients can receive "high quality ambulatory care," then 25 SLE-specific conditions or complications can be potentially prevented.[4] Phase three of the project is forthcoming and is supposed to provide more actionable solutions to move lupus care forward. For ongoing information about the ALPHA project, please see https://www.lupus.org/alpha-project-publication-and-articles.

Patient Involvement and Coping Strategies

I am encouraged that researchers are beginning to pay more attention to lupus patients and even ask them what works in coping with this life-long disease. A recent article surveyed lupus patients regarding their coping strategies, and five themes emerged: positive attitudes, social support, trust in medical treatments, healthy habits, and avoiding stress.[5] I'm glad lupus patients can participate in such projects. I hope I am doing my part as a patient/advocate by researching and distributing information to others about the devastating symptom of cognitive dysfunction in lupus. I implore you also to examine what your role should be in fighting autoimmune disease.

Current Treatments and Their Limitations

One lupus researcher stated, "there is little question that our colleagues in rheumatology who treat patients with rheumatoid or psoriatic arthritis have outdone us 'lupologists' in bringing new therapies to the community."[6] In line with this thought, they identified 10 lessons that lupus researchers can learn from RA researchers about drug development. Although the paucity of treatment options for lupus does and should anger me, it also motivates me to work for a better future.[7] Benlysta (introduced in 2011) was a good start and appears to have a good safety profile.[8] However, it is not the answer for every lupus patient, and it still has significant side effects. In my life, it has been helpful but not nearly enough, and the side effects were debilitating. As a reminder, Benlysta (belimumab) targets the B cells, which have been found to be overactive in lupus, causing the production of antibodies.[9] As mentioned in a previous section, two other medications have become available in 2021 (voclosporin & anifrolumab). Anifrolumab can improve skin and joint disease in lupus patients, but adverse events included respiratory tract infections, bronchitis, and herpes zoster, in addition to infusion reactions.[10] Afimetoran, a medication in clinical trials, binds to target TLR7/8 and inhibits them, which would help prevent overactivation of the immune system in lupus.[11] We obviously need many more targeted medications **with good safety profiles**.[12] Current lupus treatments "are associated with significant toxicity and poor long-term outcomes."[3] There are concerns about infection and malignancy with biological therapies.[13] Such severe side effects are completely unacceptable, as patients are often forced to choose no treatment or a dangerous treatment whose long-term side effects are often as bad or worse than lupus (such as cancer). Ideally, we need healthier treatments and/or healthy lifestyle changes that mitigate symptoms and reverse disease.

The Importance of Remission and Reducing Damage

The main goal, and the one that lowers organ damage accrual, is either remission of lupus or at least low disease activity.[14] Speaking of damage accrual, "prevention of damage accrual is a key objective in the management of patients with lupus."[15] Damage accrual is linked to mortality and is attributable to both the disease of lupus and the medications used, particularly corticosteroids.[15] Corticosteroid use has even been positively associated with neuropsychiatric events.[16] Steroid use has also been linked to an alteration in the microbiome.[17] In fact, "it is now a well-established fact that glucocorticoids are a major cause of irreversible damage."[18] Thus, we are back again to needing effective medications that are not damaging in terms of long-term side effects. One encouraging fact is that antimalarials (like Plaquenil), as well as Benlysta, have recently been found to be protective against damage accrual, yet these meds still have significant side effects.[15] However, there are currently no approved medications that can cure lupus or even consistently provide long-term remission.[19]

Future of Lupus Treatments

A recent survey of 500 SLE patients found that over half needed better treatments that reduced flares and symptoms and kept lupus activity low.[20] A cohort of 670 moderate-to-severe lupus patients were followed twice a year for three years, and the researchers concluded that SLE treatments are often unable to reach goals of decreased glucocorticoid use and decreased disease activity, citing a need for new treatment options.[21] Are there more treatments forthcoming? Certain researchers believe the background work has been done now, so there will be "significant beneficial changes in lupus therapeutics in the coming years."[7] As just mentioned, two other lupus medications were approved in 2021. We can also take comfort in the fact that approximately 74 targeted therapies are in clinical development.[22] "Over 30 companies are currently investing in lupus clinical trials."[3] Developing new drugs has various targets: "inflammatory cytokines, chemokines, B cells, plasma cells, intracellular signaling pathways, B/T cells co-stimulation molecules, interferons, plasmacytoid dendritic cells," etc.[22] New biologics targeting B cells, T cells, or cytokines (interferon) are currently undergoing clinical trials as potential lupus treatments.[23] Drugs that enhance Treg cells using cytokines Il-2, IL-33, & IL-6 are being studied as promising treatment candidates.[24] In fact, "the effect of biologic therapies on cognitive function appears promising," particularly for medications that target IL-1, IL-6, B cell, T cell, and JAK inhibitors.[13] Very recently, it was found that both serum IL-6 and CSF IL-6 are involved in the pathogenesis of NP+SLE, and certain monoclonal antibodies to IL-6 receptors may be effective as a future treatment.[25] Benlysta (previously mentioned) is one example of a type of monoclonal antibody medication.

Another such drug, anifrolumab, attempts to inhibit interferon, as "the majority of lupus patients are gene signature positive."[26] A recent study linking lupus and macular degeneration suggested there may be a way to target both NLRC4 & NLRP3 inflammasomes (important defense agents of the immune system, which overreact and cause inflammation in lupus), with clinical trials coming to test this hypothesis.[27] Chimeric antigen receptor therapy is another avenue researchers are exploring. It is a cell-based therapy utilizing engineered T-cells to suppress autoreactive immune cells like B cells.[28] What if intestinal barrier permeability is verified by research as a factor to lupus development? In that case, peptide drugs such as AT-1001 (Larazotide acetate)–a zonulin inhibitor studied in celiac disease–may be explored as a treatment.[29,30] As a patient, what do we do with technical information on new medications when we may not understand their biochemistry and immunology? We also do not know the potential long-term damage caused by these new medications. We must stay as informed as possible by reading about clinical trials and treatment options, discussing what we read with our rheumatologists and fellow patients, and advocating relentlessly for funding and legislative support. **Our lives depend on it!**

The Missing Patient Perspective in Guidelines

Recently, a paper stating the ten most important challenges in managing SLE was published. While I agree that all ten areas are important and make excellent goals, only one considered the patient's needs from his/her perspective, which is called "improving the network of care."[31] Similarly, an extensive review of clinical practice guidelines for lupus care found that the input of patients was missing.[32] I wonder what the top ten needs would be according to actual people with lupus. **For me, relieving daily fatigue, pain, and fever would be three on the list, but my top need would be preventing and reversing cognitive dysfunction because my brain affects everything that I do (home, work, life), and I need to be able to trust it.** As Dr. Harrison stated, "cognition is understanding the world around us, taking in information, storing it, processing it, and reacting to it."[33] Thus, cognition is incredibly broad and affects everything we do in life. People with mild cognitive impairment "complain of cognitive decline impairments which interfered with their daily activity," yet "in the majority of patients they are subclinical and mainly ignored in medical evaluation."[34] In fact, "cognitive dysfunction is still grossly under-recognized by rheumatologists,"[34] However, another study states that "it is a very common problem in people who have SLE."[35] Why is cognitive impairment under-recognized by physicians, even though it has been shown to commonly occur in lupus patients?[36] In fact, "lupus patients have identified CD (cognitive dysfunction) as one of the most distressing symptoms of their disease."[37] "Patients often report cognitive impairment as **the** most bothersome disease-related manifestation, with a great effect on their quality of life."[37] Why, then, is this major, daily, life-altering issue being ignored? As patients, we must not accept it. Cognitive dysfunction in SLE is common and associated with significant morbidity but is currently underdetected.[38] Research confirms that "an important component of medical care for the SLE patient is a thorough assessment for neurocognitive complications followed by treatment or referral to specialists in neurology, psychiatry, and psychotherapy for any neuropsychological deficits noted."[13] We must push additional researchers to discover better and more efficient ways to identify, treat, and prevent cognitive deficits in autoimmune diseases. The routine screening and evaluation of cognitive function must become a part of the SLE patient care system, since even mild impairments can interfere with quality of life and employment.[39] "There is an increasing need to conduct more research on this topic to emphasize the importance of early recognition and initiation of treatment of cognitive dysfunction, with an aim to improve quality of life for these SLE patients."[40]

Looking Ahead: Hope for Cognitive Care

Fortunately, there appears to be some hope on the horizon. Drs. Touma and Barraclough ("Lupus Foundation," 2021) are now researching the causes of cognitive dysfunction in lupus. Dr.

Touma stated, "because of the support of the Lupus Foundation of America and Lupus Canada, we will be able to improve the assessment of cognitive function in people with lupus with further research in this critical area to ultimately help with future clinical trials and improve the lives of those living with the disease." When two large organizations such as these work together for the good of patients with lupus, we can hope new information on the cause(s) and treatment of cognitive impairment will be forthcoming. Masley, although not referring specifically to cognitive changes in autoimmune disease, makes the bold claim that "we really can prevent nearly two-thirds of all dementias" (p.12) by making daily lifestyle changes.[41] Might this apply to autoimmune disease, as well? Additional research is needed to determine the answer and to discover what those optimal daily choices might be. This is highly important work. Let's get it done.

Epilogue

Regarding Masley's quote in the previous section, I have personally experienced very positive results due to lifestyle changes. In addition to researching cognitive dysfunction, I have been researching nutrition for a few years. Although there are many valuable opinions on optimal nutrition, I was initially most influenced by Dr. Mark Hyman, who wrote *Food: What the Heck Should I Eat?* I was so impressed that I went so far as to make a trip to his Ultrawellness Center in Massachusetts. There, I saw his associates: an MD also trained in functional medicine, a PA, and a dietician. Following their recommendations for diet and supplements specific to my symptoms and test results (which included genetic variations, leaky gut/dysbiosis, as well as identifying specific food sensitivities), I dramatically changed my diet and lifestyle, including adding nutritional supplements and getting off of several medications.

I changed my diet by doing an elimination diet (eliminating foods I was sensitive to) and completely cutting out gluten, dairy, and nightshades. I also dramatically reduced my sugar consumption and started buying organic, real food (not processed). Within two months, my cognition had dramatically improved. I can't describe to you how life-changing and encouraging this change was! In the past, as previously mentioned, I had seen some improvement with various medications (such as Benlysta), but the negative cognitive symptoms would always return. My cognitive impairment symptoms have been virtually gone now for two years (except when I have a migraine). I believe this change (of normal cognition) is permanent, as long as I continue to eat healthy and practice a healthy lifestyle of exercise, brain stimulation, quality sleep, and stress management. I pray that such positive changes can begin to provide hope for all patients who suffer daily from cognitive dysfunction. You may have to give up junk food, but if you can begin to trust your brain again, it is completely worth it. Surely you can deny yourself a pizza or a bagel to regain brain function! (As a bonus, my daily stomach pain resolved and my joint pain and fatigue have decreased, as well, due to diet and lifestyle changes.)

My cognition further improved when my functional medicine doctor introduced me to a new treatment: Thymosin Alpha and Thymosin Beta4 peptides. I used injectable peptides for a couple of years, and when I couldn't get them anymore, I switched to capsules. Peptides further decreased my fatigue, improved mental clarity and stamina, and decreased pain and stiffness. I have recently also weaned off hydroxychloroquine (a medication I've been taking for 17 years) since my lupus is doing so much better on the peptides. Peptides (small bits of proteins) are even more impressive due to their complete lack of side effects–something I never experienced with conventional medications like Methotrexate and Benlysta. Peptide treatment has been found to significantly delay the rise in cytokines, and repeated therapy greatly improved the survival rate of mice with SLE.[1] I am thrilled peptides are beginning to be investigated for human lupus treatment since they are cost-effective and have low toxicity.[2] They mimic normal bodily processes and help to induce immune tolerance (so the body can appropriately discriminate self from non-self). Since the immune systems of people with lupus (and other autoimmune diseases) attack their own healthy tissue, peptides can help normalize the immune system's response.

Scientists from Northwestern Medicine have successfully utilized nontoxic synthetic peptides to stimulate regulatory T-cells and block autoantibody production in lupus patients.[3] Peptide treatment can decrease inflammation and halt autoimmunity, leading to sustained remission or possibly even prevention of lupus in susceptible individuals (see review by Datta, 2021). Clinical trials are needed, but researchers hope this type of treatment can put lupus patients into permanent remission and may eventually provide a vaccine for at-risk patients.[3] Although peptide treatment currently remains controversial (not FDA approved), I am incredibly grateful for what it has done to improve my own health. I hope that I (and other lupus and arthritis sufferers) will be able to receive this nontoxic treatment consistently in the future.

I have finally found a solution to my brain fog, with diet and lifestyle, along with innovative treatments. I sincerely hope you find yours, too. Maybe something in this book will lead you to part of your answer. If I can help in any way, please contact me. Also, if you find additional resources that are helpful in the fight for normalized cognition, please contact me so I may include them in a later edition of this book. Please help me to help other lupus/autoimmune warriors. Together, we can work to eliminate this terrible symptom and, one day, these horrible autoimmune diseases.

Personal Success Factors

After completing all the research for this project and being a bit of a guinea pig myself, I believe the key to surviving and even thriving through cognitive dysfunction in lupus involves several factors. Unfortunately, I do not believe it is an easy fix, requiring the change of only one thing, but a more complex, multifactorial approach is warranted, similar to the one put forth by Dr. Bredesen for Alzheimer's disease. Even conventional medicine doctors are realizing the need for these multidomain interventions for cognitive impairment due to RCTs (randomized controlled clinical trials) like MAPT and FINGER.[4] Since we know that lupus (and other autoimmune diseases) is a complex, multifactorial disease, doesn't it make sense that it would also require a complex treatment?

The first factor for me is the need for connection and support. Personally, that's a daily connection to God through prayer, as well as consistent support from my family and friends. Second, to achieve optimal cognition, you must work to challenge your brain every day. That may mean studying a new language, reading something difficult, learning a new hobby, or doing daily BrainHQ or DuoLingo computer exercises. If it's easy for you to complete, then it's not working well to challenge your brain, and you must find something harder to learn and think about. Remember: the brain changes with new tasks. (For an excellent introduction to brain plasticity and why you want it, read Doidge's (2007) book). Third, as stated previously, it is imperative you get quality anti-inflammatory nutrition specific to *your* needs. This factor made the biggest difference in my own cognition. I highly recommend you read some of the books mentioned previously (see books by Hyman, O'Bryan, & Bredesen) or have a family member read along with you for a thorough guide to optimal nutrition for people with autoimmune disease. Suppose you don't get sufficient results with your own dietary changes. In that case, you may need to work with a functional medicine doctor and/or a functional medicine health coach to personalize your approach to eating, and add in nutritional supplements when necessary. (Note: One unfortunate issue with functional medicine testing & supplementation is the fact that costs are usually not covered by insurance, so be aware of this.)

Fourth, I have found I need downtime (relaxing, unwinding, connecting to nature)—don't discount time off as frivolous or wasteful—it is important! The fifth factor is frequent gentle exercise (that's yoga, water aerobics, walking, Zumba, light weights, and Pilates for me). My frequency for exercise is 5-6 days a week, but the type requires a lot of modification and grace based on how I'm feeling that particular day. I've learned overdoing it is just as bad as not doing it. One study showed you can reap benefits to episodic memory (memory for events) with consistently exercising thrice weekly.[5] The sixth factor is lots and lots of sleep. It took me a long time to accept that, due to my autoimmune disease, I need more sleep than the average person. I had to learn to value my extra sleep and to not be embarrassed by it in our fast-paced, never-waste-time culture. Seventh, and finally, you need to regularly see your doctors

Epilogue

and monitor your progress through laboratory tests. This has worked best for my health with a team approach: a conventional rheumatologist, a functional medicine doctor, and a dietician.

To begin to explore research on functional medicine, I highly recommend reading the paper by Beidelschies, et al.,[6] titled "Association of the Functional Medicine Model of Care with Patient-Reported Health-Related Quality-of-Life Outcomes" published in the Journal of American Medical Association. To obtain additional pain relief and maintain bodily structure, I also regularly see a chiropractor. Keeping track of all the above factors is time-consuming and not easy—they often require sacrifice and self-discipline. Sometimes, I joke that it's a part-time job, just keeping up with my health. When you begin to become yourself again, however, you will realize you have made the right choices. I challenge you to commit to investing three months in such a program. I promise it can be life-changing in the most positive way! As I end this book, I hope you have felt my writing come from a place of compassion and hope. I truly want **you** to be better, healthier, and think **clearly** about your future.

Glossary

ACE (angiotensin-converting enzyme) inhibitor: A medication used to treat high blood pressure and heart failure by relaxing blood vessels.

Adenine Dinucleotide (NAD⁺): A coenzyme involved in redox reactions and cellular energy production.

Anellovirus: A group of viruses commonly found in humans with uncertain clinical significance.

Amyloid Beta Plaques: Protein clumps found in the brains of individuals with Alzheimer's disease.

Ankylosing Spondylitis: A type of arthritis that affects the spine and sacroiliac joints.

Antiphospholipid Antibodies: Autoantibodies associated with increased risk of blood clots.

Anti-Ribosomal P Antibodies: Autoantibodies often present in patients with systemic lupus erythematosus (SLE).

Archaea: Microorganisms similar to bacteria but genetically distinct, often found in extreme environments.

Astrocytes: Star-shaped glial cells in the brain and spinal cord involved in support and repair.

Autoantigens: Normal proteins or molecules in the body that the immune system mistakenly targets.

B-cells, T-cells, Dendritic Cells: Key immune cells that detect, respond to, and remember pathogens.

Bacteroidetes: A group of bacteria commonly found in the human gut, playing roles in digestion.

B Cell-Activating Factor: A protein involved in B cell survival and autoimmune diseases.

Brain parenchyma: The functional tissue in the brain composed of neurons and glial cells.

Brodmann area 41: A region in the auditory cortex

involved in processing sound.

CD8+ T cells: Immune cells that kill infected or cancerous cells.

Choroid Plexus: Brain structure that produces cerebrospinal fluid.

Complement Cascade: A series of immune reactions leading to the destruction of pathogens.

Complement-Fixing Immune Complex: Antibody-antigen complexes that activate the complement system.

Curcumin: An anti-inflammatory compound found in turmeric.

CVA (Cerebrovascular Accident): Commonly known as a stroke.

Cyclophosphamide: A chemotherapy drug also used to suppress the immune system.

Cytokines: Small proteins important in cell signaling, especially in immune responses.

Dendritic Cells: Antigen-presenting immune cells that activate T cells.

Dentate Gyrus: A part of the hippocampus involved in memory formation.

Epigenetic Alterations: Changes in gene expression without altering DNA sequence.

Enterococcus Gallinarum: A gut bacterium linked to autoimmune conditions.

Fibromyalgia: A disorder characterized by widespread musculoskeletal pain and fatigue.

Fisetin: A plant polyphenol with antioxidant and anti-inflammatory properties.

Firmicutes: A group of gut bacteria associated with energy absorption and metabolism.

Flavones and Anthocyanins: Plant compounds with antioxidant effects.

Forkhead box P3 (FOXP3): A gene crucial for the development of regulatory T cells.

Gamma Neural Activity: Brainwave pattern associated with high-level cognitive functions.

Glycine: An amino acid that acts as a neurotransmitter in the central nervous system.

Glucocorticoids (Steroids): Hormones that reduce inflammation and suppress the immune system.

Heschl's Gyrus: Brain region involved in the perception of sound.

IL-6 (Interleukin-6): A cytokine involved in inflammation and immune response.

Immune Complexes: Antigen-antibody complexes formed during immune responses.

Immunosenescence: Aging-related decline in immune function.

Insula: Brain region involved in emotion, perception, and self-awareness.

Interleukin-1 Beta: A pro-inflammatory cytokine produced in response to infections.

Isoform Expression: The production of different forms of proteins from the same gene.

JAK/STAT3: A signaling pathway important in immunity and inflammation.

KYN/TRP (kynurenine/tryptophan): Metabolites involved in inflammation and brain function.

LLN (Left Lentiform Nucleus): A brain structure involved in movement and cognition.

Lumbar Punctures: A procedure to collect cerebrospinal fluid for testing.

Lupologist: A specialist in the study and treatment of lupus.

Melatonin: A hormone that regulates sleep-wake cycles.

Micro-Thrombi: Small blood clots that can obstruct circulation.

MHC class II and IRF5 regions: Genetic areas linked to immune function and autoimmune diseases.

MMSE (Mini-Mental State Examination): A test used to assess cognitive function.

MoCA (Montreal Cognitive Assessment): A short assessment for mild cognitive impairment.

Monocytes: White blood cells that become macrophages and digest pathogens.

Mucocutaneous Lesions: Sores or ulcers affecting the skin and mucous membranes.

Myalgia: Muscle pain.

N-methyl-D-aspartate (NMDA): A receptor for a neurotransmitter involved in learning and memory.

NETS (Neutrophil Extracellular Traps): Web-like structures expelled by neutrophils to trap pathogens.

Nicotinamide Mononucleotide (NMN): A precursor to NAD$^+$ involved in energy metabolism.

Neuroprotective: Having the ability to protect nerve cells from damage.

Neurotrophic: Promoting the survival and growth of neurons.

Oral Microbiota Dysbiosis: Imbalance of bacteria in the mouth linked to disease.

Parahippocampal Region: Brain area important for memory encoding and retrieval.

PGA (Physician Global Assessment): A tool used to assess overall disease activity.

Phenotypes: Observable traits or characteristics of an organism.

Piperlongumine: A natural compound with potential anticancer properties.

Posterior cingulate gyrus: A brain region involved in memory and emotion.

Proteinuria: Presence of excess proteins in the urine, often indicating kidney damage.

QA/KA (quinolinic acid/kynurenic acid):

Neuroactive metabolites in the kynurenine pathway.

Qigong: A traditional Chinese practice involving movement, breathing, and meditation.

Quercetin: A plant flavonoid with antioxidant and anti-inflammatory effects.

Reactive oxygen species (ROS): Chemically reactive molecules involved in cell signaling and damage.

Ro60 commensals: Microorganisms linked with the Ro60 autoantigen in autoimmune diseases.

Ruminococcus gnavus: A gut bacterium associated with inflammatory bowel disease.

S100B: A protein used as a marker of brain injury.

SARS-CoV-2 infection: The viral infection that causes COVID-19.

Serositis: Inflammation of the lining around organs, such as the lungs or heart.

single nucleotide polymorphisms (SNPs): Variations in a single DNA building block.

SLEDAI (Systemic Lupus Erythematosus Disease Activity Index): A scoring system for lupus disease activity.

SPECT (Single Photon Emission Computed Tomography): A brain imaging technique.

Spatial Navigation Task: A test used to assess spatial memory and navigation skills.

Synaptic Pruning: The process of eliminating extra neurons and synapses during development.

Synovial Tissue: Tissue that lines joints and produces lubricating fluid.

Thymosin beta 4 peptide: A molecule involved in tissue regeneration and immune modulation.

Thrombotic and nonthrombotic: Related to clotting and non-clotting conditions.

Treg/Th17/Th1 imbalance: A skewed ratio of T-cell

types that contributes to autoimmune diseases.

Treg cells (Regulatory T cells): Cells that suppress immune responses and maintain tolerance.

Transcranial Neuromodulation: A technique to modulate brain activity using noninvasive stimulation.

TSPO protein: A protein used as a marker for inflammation in the brain.

Ulcerative colitis: A chronic inflammatory bowel disease affecting the colon.

Vasogenic edema: Swelling caused by fluid leakage from blood vessels into brain tissue.

Ventromedial prefrontal cortex and orbitofrontal cortex: Brain areas involved in decision-making and emotional regulation.

Vermis: A central part of the cerebellum involved in posture and locomotion.

Voxel-mirrored homotopic connectivity: A measure of brain connectivity between hemispheres.

Voxels: 3D pixels used in brain imaging scans.

Wernicke's aphasia: A language disorder caused by damage to the brain's language-processing areas.

Yin Yoga: A slow-paced style of yoga with postures held for longer periods.

References

Introduction

[1] Tamilou F, Arnaud L, Talarico R, et al. Systemic lupus erythematosus: state of the art of clinical practice guidelines. *RMD Open.* 2018;4(1):e000793. doi:10.1136/rmdopen-2018-000793

[2] Knight JS, Caricchio R, Casanova JL, et al. The intersection of COVID-19 and autoimmunity. *J Clin Invest.* 2021;131(24):e154886. doi:10.1172/JCI154886

[3] Apple AC, Peluso MJ, Asken BM, et al. Risk factors and abnormal cerebrospinal fluid associate with cognitive symptoms after mild COVID-19. *Ann Clin Transl Neurol.* 2022;doi:10.1002/acn3.51498

[4] Li BZ, Zhou HY, Guo B, et al. Dysbiosis of oral microbiota is associated with systemic lupus erythematosus. *Arch Oral Biol.* 2020;113:104708. doi:10.1016/j.archoralbio.2020.104708

[5] Boumpas DT. Management of neuropsychiatric SLE. *Lupus Sci Med.* 2022;9(Suppl 1):A8.2. Available from: https://lupus.bmj.com/content/9/Suppl_1/A8.2

Chapter 1

[1] Thomas DE. *The Lupus Encyclopedia: A Comprehensive Guide for Patients and Families. Baltimore, MD:* Johns Hopkins University Press; 2014.

[2] Lopez P, de Paz B, Rodriquez-Carrio J, et al. Th17 responses and natural IgM antibodies are related to gut microbiota composition in systemic lupus erythematosus patients. *Sci Rep.* 2016;6:240-72. doi:10.1038/srep24072

[3] Tamilou F, Arnaud L, Talarico R, et al. Systemic lupus erythematosus: state of the art of clinical practice guidelines. *RMD Open*. 2018;4(1):e000793. doi:10.1136/rmdopen-2018-000793

[4] Wallace DJ, Hahn BH. Dubois' *Lupus Erythematosus*. 7th ed. Philadelphia, PA: Lippincott Williams & Wilkins; 2007.

[5] Diamond B, Volpe BT. A model for lupus brain disease. *Immunol Rev*. 2012;248(1):56-67. doi:10.1111/j.1600-065X.2012.01137.x

[6] Atkinson J, Yu CY. Complement in SLE. *BMJ*. 2021;8(1). doi:10.1186/ar586

[7] Yuen H, Cunningham M. Optimal management of fatigue in patients with systemic lupus erythematosus: a systematic review. *Ther Clin Risk Manag*. 2014;10:775-786. doi:10.2147/TCRM.S56063

[8] Greiling TM, Dehner C, Chen X, et al. Commensal orthologs of the human autoantigen Ro60 as triggers of autoimmunity in lupus. *Sci Transl Med*. 2018;10(434). doi:10.1126/scitranslmed.aan2306

[9] Kang HK, Liu M, Datta SK. Low-dose peptide tolerance therapy of lupus generates plasmacytoid dendritic cells that cause expansion of autoantigen-specific regulatory T cells and contraction of inflammatory Th17 cells. *J Immunol*. 2007;178(12):7849-7858. doi:10.4049/jimmunol.178.12.7849

[10] Datta SK. Harnessing tolerogenic histone peptide epitopes from nucleosomes for selective down-regulation of pathogenic autoimmune response in lupus (past, present, and future). *Front Immunol*. 2021;12:629-807. doi:10.3389/fimmu.2021.629807

[11] Goh YP, Naidoo P, Ngian GS. Imaging of systemic lupus erythematosus. Part I: CNS, cardiovascular, and thoracic manifestations. *Clin Radiol*. 2013;68(2):181-191. doi:10.1016/j.crad.2012.06.110

[12] Yu C, Gershwin ME, Chang C. Diagnostic criteria for systemic lupus erythematosus: A critical review. *J Autoimmun*. 2014;48-49:10-13. doi:10.1016/j.jaut.2014.01.004

[13] Lateef A, Petri M. Unmet medical needs in systemic lupus erythematosus. *Arthritis Res Ther*. 2012;14(4):S4. doi:10.1186/ar3919

[14] Bakshi J, Segura B, Wincup C, Rahman A. Unmet needs in the pathogenesis and treatment of systemic lupus erythematosus. *Clin Rev Allergy Immunol*. 2018;55(3):352–367. doi:10.1007/s12016-017-8640-5

[15] Petri M, Orbai AM, Alarcón GS, et al. Derivation and validation of Systemic Lupus International Collaborating Clinics (SLICC) classification criteria for systemic lupus erythematosus. *Arthritis Rheumatol*. 2012;64(8):2677-2686. doi:10.1002/art.34473

[16] Aringer M, Costenbader KH, Daikh DI, et al. 2019 EULAR/ACR classification criteria for systemic lupus erythematosus. *Arthritis Rheumatol*. 2019;71(9):1400-1412. doi:10.1002/art.40930

[17] Li BZ, Zhou HY, Guo B, et al. Dysbiosis of oral microbiota is associated with systemic lupus erythematosus. *Arch Oral Biol*. 2020;113:104708. doi:10.1016/j.archoralbio.2020.104708

[18] Eudy AM, Rogers JL, Corneli A, et al. Intermittent and persistent Type 2 lupus: patient perspectives on two distinct patterns of Type 2 SLE symptoms. *Lupus Sci Med*. 2022;9(1):e000705. doi:10.1136/lupus-2022-000705

[19] Banchereau R, Hong S, Cantarel B, et al. Personalized immunomonitoring uncovers molecular networks that stratify lupus patients. *Cell*. 2016;165(3):551-565. doi:10.1016/j.cell.2016.05.057

[20] Montjoye S, Bolan B, Van Raemdonck J, Houssiau F. Very late-onset systemic lupus erythematosus as unusual cause of reversible functional and cognitive impairments in an octogenarian patient. *Eur J Case Rep Intern Med*. 2020;7(8):001570. doi:10.12890/2020_001570

[21] Magro-Checa C, Zirkzee EJ, Huizinga TW, Steup-Beekman GM. Management of neuropsychiatric systemic lupus erythematosus: current

approaches and future perspectives. *Drugs*. 2016;76(4):459-483. doi:10.1007/s40265-015-0534-3

[22] Liang H, Tian X, Lan-Yu C, Yan-Yan C, Wang C. Effect of psychological intervention on health-related quality of life in people with systemic lupus erythematosus: a systematic review. *Int J Nurs Sci*. 2014;2(3):298-305. doi:10.1016/j.ijnss.2014.07.008

[23] Nicolaou O, Kousios A, Hadisavvas A, Lauwerys B, Sokratous K, Kyriacou K. Biomarkers of systemic lupus erythematosus identified using mass spectrometry-based proteomics: a systematic review. *J Cell Mol Med*. 2017;21(5):993-1012. doi:10.1111/jcmm.13031

[24] Kapadia M, Bijjelic D, Zhao H, Donglai M, et al. Effects of sustained i.c.v. infusion of lupus CSF and autoantibodies on behavioral phenotype and neuronal calcium signaling. *Acta Neuropathol Commun*. 2017;5:47. doi:10.1186/s40478-017-0473-1

[25] Barber MR, Drenkard C, Falasinnu T, et al. Global epidemiology of systemic lupus erythematosus. *Nat Rev Rheumatol*. 2021;17(9):515-532. doi:10.1038/s41584-021-00668-1

[26] Jafri K, Patterson S, Lanata C. Central nervous system manifestations of systemic lupus erythematosus. *Rheum Dis Clin North Am*. 2017;43(4):531–545. doi:10.1016/j.rdc.2017.06.003

[27] Tse K, Sangodkar S, Bloch L, et al. The ALPHA Project: Establishing consensus and prioritization of global community recommendations to address major challenges in lupus diagnosis, care, treatment and research. *Lupus Sci Med*. 2020;8(1):e000433. doi:10.1136/lupus-2020-000433

[28] Schlencker A, Messer L, Ardizzone M, et al. Improving patient pathways for systemic lupus erythematosus: a multistakeholder pathway optimisation study. *Lupus Sci Med*. 2022;9(1):e000700. doi:10.1136/lupus-2022-000700

[29] Barber MR, Drenkard C, Falasinnu T, et al. Global epidemiology of systemic lupus erythematosus. *Nat Rev Rheumatol*. 2021;17(9):515-532. doi:10.1038/s41584-021-00668-1

[30] Bendorius M, Po C, Jeltsch-David H. From systemic inflammation to neuroinflammation: the case of neurolupus. *Int J Mol Sci*. 2018;19(11):3588. doi:10.3390/ijms19113588

[31] Abend AH, He I, Bahroos N, et al. Estimation of prevalence of autoimmune diseases in the United States using electronic health record data. *J Clin Invest*. 2024;135(4):e178722. doi:10.1172/JCI178722

[32] Izmirly PM, Ferucci ED, Somers EC, et al. Incidence rates of systemic lupus erythematosus in the USA: estimated from a meta-analysis of the Centers for Disease Control and Prevention national lupus registries. *BMJ*. 2021;doi:10.1136/lupus-2021-000614

[33] Gergianaki I, Bertsias G. Systemic lupus erythematosus in primary care: An update and practical messages for the general practitioner. *Front Med*. 2018;5:161. doi:10.3389/fmed.2018.00161

[34] Carlomagno S, Migliaresi S, Ambrosone I, et al. Cognitive impairment in systemic lupus erythematosus: a follow-up study. *J Neurol*. 2000;247(4):273–279. doi:10.1007/s004150050583

[35] Somers EC, Marder W, Cagnoli P, et al. Population-based incidence and prevalence of systemic lupus erythematosus. *Arthritis Rheumatol*. 2014;66(2):369-378. doi:10.1002/art.38238

[36] Magro-Checa C, Zirkzee EJ, Huizinga TW, Steuo-Beekman GM. Management of neuropsychiatric systemic lupus erythematosus: Current approaches and future perspectives. *Drugs*. 2016;76:459-483. doi:10.1007/s40265-015-0534-3

[37] Qian X, Ji F, Ng KK, et al. Brain white matter extracellular free-water increases are related to reduced neurocognitive function in systemic lupus erythematosus. *Rheumatology (Oxford)*. 2021 Jun 22:511. doi:10.1093/rheumatology/keab511

[38] Rees F, Doherty M, Grainge MJ, Lanyon P, Zhang W. The worldwide incidence and prevalence of systemic lupus erythematosus: A systematic review of epidemiological studies. *Rheumatology (Oxford)*. 2017;56(11):1945-1961. doi:10.1093/rheumatology/kex260

[39] Greiling T, Dehner C, Chen X, et al. Commensal orthologs of the human autoantigen Ro60 as triggers of autoimmunity in lupus. *Science Transl Med*. 2018;10(434):eaan2306. doi:10.1126/scitranslmed.aan2306

[40] Fava A, Petri M. Systemic lupus erythematosus: Diagnosis and clinical management. *J Autoimmun*. 2019;96:1-13. doi:10.1016/j.jaut.2018.11.001

[41] Plantinga L, Lim S, Bowling C, Drenkard C. Perceived stress and reported cognitive symptoms among Georgia patients with systemic lupus erythematosus. *Lupus*. 2017;26(10):1064-1071. doi:10.1177/0961203317693095

[42] Cuffari B. What is neuropsychiatric lupus (NPSLE)? *News-Medical.net*. Published March 2021. Accessed April 15, 2025. https://www.news-medical.net

Chapter 2

[1] Liang H, Tian X, Cao LY, Chen YY, Wang CM. Effect of psychological intervention on health-related quality of life in people with systemic lupus erythematosus: a systematic review. *Int J Nurs Sci*. 2014;1(3):298-305. doi:10.1016/j.ijnss.2014.07.008

[2] Bendorius M, Po C, Muller S, Jeltsch-David H. From systemic inflammation to neuroinflammation: the case of neurolupus. *Int J Mol Sci*. 2018;19(11):3588. doi:10.3390/ijms19113588

[3] Grabich S, Farrelly E, Ortmann R, Pollack M, Wu SSJ. Real-world burden of systemic lupus erythematosus in the USA: a comparative cohort study from the Medical Expenditure Panel Survey (MEPS)

2016–2018. *Lupus Sci Med.* 2022;9(1):e000640. doi:10.1136/lupus-2021-000640

[4] Fanouriakis A, Bertsias G. Changing paradigms in the treatment of systemic lupus erythematosus. *Lupus Sci Med.* 2019;6(1):e000310. doi:10.1136/lupus-2018-000310

[5] Celhar T, Fairhurst A. Modelling clinical systemic lupus erythematosus: similarities, differences and success stories. *Rheumatology.* 2017;56(1):i88-i99. doi:10.1093/rheumatology/kew400

[6] Gergianaki I, Bertsias G. Systemic lupus erythematosus in primary care: an update and practical messages for the general practitioner. *Front Med (Lausanne).* 2018;5:161. doi:10.3389/fmed.2018.00161

[7] Hanly JG, Gordon C, Bae S, et al. Neuropsychiatric events in systemic lupus erythematosus: predictors of occurrence and resolution in a longitudinal analysis of an international inception cohort. *Arthritis Rheumatol.* 2021. doi:10.1002/art.41876

[8] Urowitz M. 02 Hydroxychloroquine myopathy: cardiac and skeletal muscle toxicity. *Lupus Sci Med.* 2021;9. doi:10.1136/lupus-2021-la.2

[9] Sood A, Raji M. Cognitive impairment in elderly patients with rheumatic disease and the effect of disease-modifying anti-rheumatic drugs. *Clin Rheumatol.* 2021. doi:10.1007/s10067-020-05372-1

[10] Popescu A, Kao AH. Neuropsychiatric systemic lupus erythematosus. *Curr Neuropharmacol.* 2011;9(3):449-457. doi:10.2174/157015911796557984

[11] Bakshi J, Segura B, Wincup C, Rahman A. Unmet needs in the pathogenesis and treatment of systemic lupus erythematosus. *Clin Rev Allergy Immunol.* 2018;55(3):352–367. doi:10.1007/s12016-017-8640-5

[12] Figueroa-Parra G, Heien HC, Warrington KJ, et al. Treatment trends of systemic lupus erythematosus from 2007 to 2023 in the USA. *Lupus Sci Med.* 2024;11(2):e001317. doi:10.1136/lupus-2024-001317

[13] Schoenbach A. Keeping yourself healthy on immunosuppressive drugs. LupusCorner website. Published 2020. Accessed April 15, 2025. https://llupuscorner.com/immunosuppressive-drugs-strategies-for-staying-healthy

[14] Dörner T. B cells as a therapeutic target in SLE: where we are today and where we may be tomorrow. *Lupus Sci Med*. 2023;9. Accessed April 15, 2025. https://www.researchgate.net/publication/367887305_07_B_cells_as_a_therapeutic_target_in_SLE_where_we_are_today_and_where_we_may_be_tomorrow

[15] Fanouriakis A, Bertsias G. Changing paradigms in the treatment of systemic lupus erythematosus. *BMJ Lupus Sci Med*. 2018;6(1):e000310. doi:10.1136/lupus-2018-000310

[16] Venkatadri R, Sabapathy V, Dogan M, Sharma R. Targeting regulatory T cells for therapy of lupus nephritis. *Front Pharmacol*. 2021;12:806612. doi:10.3389/fphar.2021.806612

[17] Northcott M, Jones S, Koelmeyer R, et al. Type 1 interferon status in systemic lupus erythematosus: a longitudinal analysis. *Lupus Sci Med*. 2021;9(1):e000625. doi:10.1136/lupus-2021-000625

[18] Lupus Foundation of America. Lupus Foundation of America celebrates FDA approval of Saphnelo (Anifrolumab-fnia) as a new treatment for lupus. *PR Newswire*. Published July 30, 2021. Accessed April 15, 2025. https://www.prnewswire.com/news-releases/lupus-foundation-of-america-celebrates-fda-approval-of-saphnelo-anifrolumab-fnia-as-a-new-treatment-for-lupus-301346041.html

[19] Masurkar PP, Reckleff J, Princic N, et al. Real-world treatment patterns, healthcare resource utilisation and costs in patients with SLE in the USA. *Lupus Sci Med*. 2024;11(2):e001290. doi:10.1136/lupus-2024-001290

[20] Thurman J, Serkova N. Non-invasive imaging to monitor lupus nephritis and neuropsychiatric systemic lupus erythematosus. *F1000Research*. 2015;4:153. doi:10.12688/f1000research.6587.2

[21] Jiang N, Li M, Zhang H, et al. Sirolimus versus tacrolimus for systemic lupus erythematosus treatment: results from a real-world CSTAR cohort study. *Lupus Sci Med.* 2021. doi:10.1136/lupus-2021-00617

[22] Fanouriakis A, Bertsias G. Changing paradigms in the treatment of systemic lupus erythematosus. *Lupus Sci Med.* 2018;6(1):e000310. doi:10.1136/lupus-2018-000310

[23] Datta SK. Harnessing tolerogenic histone peptide epitopes from nucleosomes for selective down-regulation of pathogenic autoimmune response in lupus (past, present, and future). *Front Immunol.* 2021;12:629807. doi:10.3389/fimmu.2021.629807

[24] Kello N, Anderson E, Diamond B. Cognitive dysfunction in systemic lupus erythematosus: a case for initiating trials. *Arthritis Rheumatol.* 2019;71(9):1413-1425. doi:10.1002/art.40933

[25] Van Vollenhoven RF, Bertsias G, Doria A, et al. 2021 DORIS definition of remission in SLE: final recommendations from an international task force. *Lupus Science & Medicine.* 2021;8:e000538. doi:10.1136/lupus-2021-000538

[26] Inês L. Disease activity and treatment targets in SLE. Lupus Sci Med. 2022;9:e000623. doi:10.1136/lupus-2022-la.24. Available from: https://www.researchgate.net/publication/367887554_24_Disease_activity_and_treatment_targets_in_SLE

Chapter 3

[1] Castellini-Pérez O, Iakovliev A, Martínez-Bueno M, et al. Genetic evaluation of molecular traits in systemic lupus erythematosus. *Lupus Sci Med.* 2022;9(Suppl 1):S10.1. doi:10.1136/lupus-2022-elm2022.19

[2] Khoury LE, Zarfeshani A, Diamond B. Using the mouse to model human diseases: cognitive impairment in systemic lupus erythematosus. *J Rheumatol.* 2020;47(7):1145-1149. doi:10.3899/jrheum.200410

[3] Crow Y. How interferonopathies inform SLE pathogenesis. *Lupus Sci Med.* 2021;8(Suppl 1):05. doi:10.1136/lupus-2021-la.5

[4] Gorji AE, Roudbari Z, Alizadeh A, Sadeghi B. Investigation of systemic lupus erythematosus (SLE) with integrating transcriptomics and genome wide association information. *Gene.* 2019;706:181-187. doi:10.1016/j.gene.2019.05.004

[5] Yeoh S, Dias SS, Isenberg DA. Advances in systemic lupus erythematosus. *Medicine (Baltimore).* 2017;46(2):84-92.

[6] Datta SK. Harnessing tolerogenic histone peptide epitopes from nucleosomes for selective down-regulation of pathogenic autoimmune response in lupus (past, present, and future). *Front Immunol.* 2021;12:629807. doi:10.3389/fimmu.2021.629807

[7] Wang Y, Xie X, Zhang C, et al. Rheumatoid arthritis, systemic lupus erythematosus and primary Sjögren's syndrome shared megakaryocyte expansion in peripheral blood. *Ann Rheum Dis.* 2022;81(3). doi:10.1136/annrheumdis-2021-220066

[8] Brown GJ, Cañete PF, Wang H, et al. TLR7 gain-of-function genetic variation causes human lupus. *Nature.* 2022;605(7909):349–356. doi:10.1038/s41586-022-04642-z

[9] Fava A, Petri M. Systemic lupus erythematosus: Diagnosis and clinical management. *J Autoimmun.* 2019;96:1-13. doi:10.1016/j.jaut.2018.11.001

[10] Rivas-Larrauri F, Yamazaki-Nakashimada MA. Systemic lupus erythematosus: Is it one disease? *Reumatol Clin.* 2016;12(5):274-281. doi:10.1016/j.reuma.2016.01.005

[11] Leffers HC, Lange T, Collins C, Ulff-Møller CJ, Jacobsen S. The study of interactions between genome and exposome in the development

of systemic lupus erythematosus. *Autoimmun Rev.* 2019;18(4):382-392. doi:10.1016/j.autrev.2018.11.005

[12] Ulff-Møller CJ, Simonsen J, Kyvik KO, Jacobsen S, Frisch M. Family history of systemic lupus erythematosus and risk of autoimmune disease: nationwide cohort study in Denmark 1977–2013. *Rheumatology (Oxford).* 2017;56(6):957-964. doi:10.1093/rheumatology/kex005

[13] Munroe ME, Young KA, Guthridge JM, et al. Pre-clinical autoimmunity in lupus relatives: self-reported questionnaires and immune dysregulation distinguish relatives who develop incomplete or classified lupus from clinically unaffected relatives and unaffected, unrelated individuals. *Front Immunol.* 2022;13:866181. doi:10.3389/fimmu.2022.866181

[14] Lambers W, Arends S, Roozendaal C, et al. Prevalence of systemic lupus erythematosus-related symptoms assessed by using the Connective Tissue Disease Screening Questionnaire in a large population-based cohort. *Lupus Sci Med.* 2021;8:e000555. doi:10.1136/lupus-2021-000555

[15] Cannerfelt B, Nystedt J, Jonsen A, et al. White matter lesions and brain atrophy in systemic lupus erythematosus patients: correlation to cognitive dysfunction in a cohort of systemic lupus erythematosus patients using different definition models for neuropsychiatric systemic lupus erythematosus. *Lupus.* 2018;27:1140-1149. doi:10.1177/0961203318763533

[16] Dorner T. B cells as a therapeutic target in SLE: where we are today and where we may be tomorrow. *Lupus Sci Med.* 2022;9(Suppl 1):07. doi:10.1136/lupus-2022-elm2022.7

[17] Bakshi J, Segura B, Wincup C, Rahman A. Unmet needs in the pathogenesis and treatment of systemic lupus erythematosus. *Clin Rev Allergy Immunol.* 2018;55(3):352–367. doi:10.1007/s12016-017-8640-5

[18] Schoenbach A. T cells, the immune system, and lupus. *LupusCorner.* Published 2020. Accessed April 15, 2025. https://lupuscorner.com/t-cells-immune-system-and-lupus

[19] Morris G, Berk M, Walder K, Maes M. Central pathways causing fatigue in neuro-inflammatory autoimmune illnesses. *BMC Med.* 2015;13:28. doi:10.1186/s12916-014-0259-2

[20] Okamoto H, Kobayashi A, Yamanaka H. Cytokines and chemokines in neuropsychiatric syndromes of systemic lupus erythematosus. *J Biomed Biotechnol.* 2010;2010:268436. doi:10.1155/2010/268436

[21] Venkatadri R, Sabapathy V, Dogan M, Sharma R. Targeting regulatory T cells for therapy of lupus nephritis. *Front Pharmacol.* 2021;12:806612. doi:10.3389/fphar.2021.806612

[22] Lopez P, de Paz B, Rodriquez-Carrio J, et al. Th17 responses and natural IgM antibodies are related to gut microbiota composition in systemic lupus erythematosus patients. *Sci Rep.* 2016;6:240-72. doi:10.1038/srep24072

[23] von Boehmer H, Daniel C. Therapeutic opportunities for manipulating T(Reg) cells in autoimmunity and cancer. *Nat Rev Drug Discov.* 2013;12(1):51-63. doi:10.1038/nrd3683

[24] Du J, Wang Q, Yang S, et al. FOXP3 exon 2 controls Treg stability and autoimmunity. *Sci Immunol.* 2022;7(72):eabo5407. doi:10.1126/sciimmunol.abo5407

[25] Tsokos GC. Cytokines in SLE: beyond IFN. *Lupus Sci Med.* 2022;9(Suppl 1):21. doi:10.1136/lupus-2022-la.21

[26] Kakati S, Barman B, Ahmed S, Hussain M. Neurological manifestations in systemic lupus erythematosus: A single center study from North East India. *J Clin Diagn Res.* 2017;11(1):OC05-OC09. doi:10.7860/JCDR/2017/23773.9280

[27] Aparicio-Soto M, Sánchez-Hidalgo M, Alarcón-de-la-Lastra C. An update on diet and nutritional factors in systemic lupus erythematosus management. *Nutr Res Rev.* 2017;30(1):118-137. doi:10.1017/S0954422417000026

[28] Lopez-Lopez L, Nieves-Plaza M, Castro M, et al. Mitochondrial DNA damage is associated with damage accrual and disease duration in

patients with systemic lupus erythematosus. *Lupus.* 2014;23(11):1133-1141. doi:10.1177/0961203314537769

[29] Caielli S, Cardenas J, de Jesus AA, et al. Erythroid mitochondrial retention triggers myeloid-dependent type I interferon in human SLE. *Cell.* 2021;184(17):4464-4479. doi:10.1016/j.cell.2021.07.021

[30] Huang H. A. Multicentre, randomized, double blind, parallel design, placebo controlled study to evaluate the efficacy and safety of Uthever (NMN supplement), an orally administered supplementation in middle aged and older adults. *Front Aging.* 2022;3:851698. doi:10.3389/fragi.2022.851698

[31] Gergianaki I, Bertsias G. Systemic lupus erythematosus in primary care: an update and practical messages for the general practitioner. *Front Med.* 2018;5:161. doi:10.3389/fmed.2018.00161

[32] Wallace DJ, Hahn BH. *Dubois' Lupus Erythematosus.* 7th ed. Philadelphia, PA: Lippincott Williams & Wilkins; 2007. ISBN 978-0-78-179394-0. Accessed April 15, 2025. https://www.wolterskluwer.com/en/solutions/ovid/dubois-lupus-erythematosus-3484

[33] Iwata S, Tanaka Y. Association of viral infection with the development and pathogenesis of systemic lupus erythematosus. *Front Med.* 2022;9:849120. doi:10.3389/fmed.2022.849120

[34] Magro-Checa C, Seup-Beekman G, Huizinga T, van Buchem MA, Ronen I. Laboratory and neuroimaging biomarkers in neuropsychiatric systemic lupus erythematosus: Where do we stand, where to go? *Front Med.* 2018;5:340. doi:10.3389/fmed.2018.00340

[35] Morand E. Interferon inhibition and the future management of SLE. *Lupus Sci Med.* 2021;8:6. doi:10.1136/lupus-2021-la.6

[36] Murayama G, Chiba A, Kuga T, et al. Inhibition of mTOR suppresses IFNalpha production and the STING pathway in monocytes from systemic lupus erythematosus patients. *Rheumatology.* 2020;59(10). doi:10.1093/rheumatology/keaa060

[37] Roy ER, Ghiu G, Li S, et al. Concerted type I interferon signaling in microglia and neural cells promotes memory impairment associated with amyloid B plaques. *Immunity.* 2022;55(5):798-812. doi:10.1016/j.immuni.2022.03.018

[38] Haim L, Cyzeriat K, Carrillo-de Sauvage M, et al. The JAK/STAT3 pathway is a common inducer of astrocyte reactivity in Alzheimer's and Huntington's diseases. *J Neurosci.* 2015;35(6):2817-2829. doi:10.1523/JNEUROSCI.3516-14.2015

[39] Thurman J, Serkova N. Non-invasive imaging to monitor lupus nephritis and neuropsychiatric systemic lupus erythematosus. *F1000Research.* 2015;4:153. doi:10.12688/f1000research.6587.2

[40] Lee JM, Chen MH, Chou KY, Chao Y, Chen MH, Tsai CY. Novel immunoprofiling method for diagnosing SLE and evaluating therapeutic response. *Lupus Sci Med.* 2022;9(1):e000693. doi:10.1136/lupus-2022-000693

[41] König MF. The microbiome in autoimmune rheumatic disease. *Best Pract Res Clin Rheumatol.* 2020;34:101473. doi:10.1016/j.berh.2019.101473

[42] Kono M, Nagafuchi Y, Hirofumi S, Keishi F. The impact of obesity and a high-fat diet on clinical and immunological features in systemic lupus erythematosus. *Nutrients.* 2021;13(2):504. doi:10.3390/nu13020504

[43] Tedeschi S, Barbhaiya M, Sparks J, et al. Dietary patterns and risk of systemic lupus erythematosus in women. *Lupus.* 2020;29(1):67-73. doi:10.1177/0961203319888791

[44] Katz-Agranov N, Zandman-Goddard G. The microbiome links between aging and lupus. *Autoimmun Rev.* 2021;20(3):102765. doi:10.1016/j.autrev.2021.102765

[45] Humphreys C. Intestinal permeability. In: Pizzorno JE, Murray MT, eds. *Textbook of Natural Medicine.* 5th ed. Elsevier; 2020:166-177. doi:10.1016/B978-0-323-43044-9.00019-4

[46] Fasano A. Zonulin, regulation of tight junctions, and autoimmune diseases. *Ann N Y Acad Sci.* 2012;1258(1):25-33. doi:10.1111/j.1749-6632.2012.06538.x

[47] Tan D, Konduri S, Erikci Ertunc M, et al. A class of anti-inflammatory lipids decrease with aging in the central nervous system. *Nat Chem Biol.* 2023;19:187-197. doi:10.1038/s41589-022-01165-6

[48] Chen KT, Chen YC, Fan YH, et al. Rheumatic diseases are associated with a higher risk of dementia: a nation-wide, population-based, case-control study. *Int J Rheum Dis.* 2018;21(2):373-380. doi:10.1111/1756-185X.13246

[49] Gronke K, Nguyen M, Fuhrmann H, Santamaria de Souza N, Schumacher J, et al. Translocating gut pathobiont *Enterococcus gallinarum* induces TH17 and IgG3 anti-RNA-directed autoimmunity in mouse and human. *Sci Transl Med.* 2025;17(784):eadj6294. doi:10.1126/scitranslmed.adj6294

[50] Shi S, Zhang Q, Sang Y, et al. Probiotic *Bifidobacterium longum* BB68S improves cognitive functions in healthy older adults: a randomized, double-blind, placebo-controlled trial. *Nutrients.* 2023;15(1):51. doi:10.3390/nu15010051

[51] Silverman GJ, Cornwell M, Izmirly P, et al. 1103 Lupus clinical flares in patients with gut pathobiont blooms share a novel peripheral blood transcriptomic immune activation profile. *Lupus Sci Med.* 2022;9. doi:10.1136/lupus-2022-lupus21century.68

[52] Li BZ, Zhou HY, Guo B, et al. Dysbiosis of oral microbiota is associated with systemic lupus erythematosus. *Arch Oral Biol.* 2020;113:104708. doi:10.1016/j.archoralbio.2020.104708

[53] Xiang S, Qu Y, Qian S, et al. Association between systemic lupus erythematosus and disruption of gut microbiota: a meta-analysis. *Lupus Sci Med.* 2022;9(1). doi:10.1136/lupus-2021-000599

[54] Thio HB. The microbiome in psoriasis and psoriatic arthritis: the skin perspective. *J Rheumatol Suppl.* 2018;94:30-21. doi:10.3899/jrheum.180133

[55] Greiling TM, Dehner C, Chen X, et al. Commensal orthologs of the human autoantigen Ro60 as triggers of autoimmunity in lupus. *Sci Transl Med.* 2018;10(434). doi:10.1126/scitranslmed.aan2306

[56] Zhu S, Jiang Y, Xu K, et al. The progress of gut microbiome research related to brain disorders. *J Neuroinflammation.* 2020;17(1):25. doi:10.1186/s12974-020-1705-z

[57] Rossato S, Oakes EG, Barbhaiya M, et al. Ultraprocessed food intake and risk of systemic lupus erythematosus among women observed in the Nurses' Health Study cohorts. *Arthritis Care Res (Hoboken).* Published online June 27, 2024. doi:10.1002/acr.25395

[58] Kalim H, Pratama M, Mahardini E, et al. Accelerated immune aging was correlated with lupus-associated brain fog in reproductive-age systemic lupus erythematosus patients. *Int J Rheum Dis.* 2020;23(5):620–626. doi:10.1111/1756-185X.13816

[59] Li W, Qin L, Feng R, et al. Emerging senolytic agents derived from natural products. *Mech Ageing Dev.* 2019;181:1-6. doi:10.1016/j.mad.2019.05.001

[60] Ramakrishnan A, et al. Cerebrospinal fluid immune dysregulation during healthy brain aging and cognitive impairment. *Cell.* 2022;185(26). doi:10.1016/j.cell.2022.11.019

Chapter 4

[1] Hanly J, Omisade A, Su L, et al. Cognitive function in systemic lupus erythematosus, rheumatoid arthritis, and multiple sclerosis assessed by computerized neuropsychological tests. *Arthritis Rheumatol.* 2010;62(5):1478-1486. doi:10.1002/art.27404

[2] Doidge N. *The Brain That Changes Itself: Stories of Personal Triumph from the Frontiers of Brain Science.* Penguin Books Ltd; 2007. https://a.co/d/e1dpLX

[3] Mackay M. Lupus brain fog: A biologic perspective on cognitive impairment, depression, and fatigue in systemic lupus erythematosus. *Immunol Res.* 2015;63:26-37. doi:10.1007/s12026-015-8716-3

[4] Griffin RM. Lupus fog and memory problems. *WebMD*. Available at: https://www.webmd.com/lupus/features/lupus-fog-memory-problems#1. Accessed April 15, 2025.

[5] Lupus Foundation of America. How lupus affects memory. *Lupus.org*. Available at: https://www.lupus.org/resources/how-lupus-affects-memory. Accessed April 15, 2025.

[6] Chandler MJ, Parks AC, Marsiske M, Rotblatt LJ, Smith GE. Everyday impact of cognitive interventions in mild cognitive impairment: A systematic review and meta-analysis. *Neuropsychol Rev.* 2017;26(3):225-251. doi:10.1007/s11065-016-9330-4

[7] Langa KM, Levine DA. The diagnosis and management of mild cognitive impairment: A clinical review. *JAMA.* 2014;312(23):2551-2561. doi:10.1001/jama.2014.13806

[8] Holthe T, Halvorsrud L, Karterud D, Hoel K, Lund A. Usability and acceptability of technology for community-dwelling older adults with mild cognitive impairment and dementia: A systematic literature review. *Clin Interv Aging.* 2018;13:863-886. doi:10.2147/CIA.S154717

[9] Mueller KD. A review of computer-based cognitive training for individuals with mild cognitive impairment and Alzheimer's disease. *Perspect ASHA Spec Interest Groups.* 2019;1(1). doi:10.1044/persp1.SIG2.47

[10] Kozora E, Ellison MC, West S. Reliability and validity of the proposed American College of Rheumatology neuropsychological battery for systemic lupus erythematosus. *Arthritis Rheumatol.* 2004;51(5):810-818. doi:10.1002/art.20692

[11] Moraes-Fontes M, Lucio I, Santos C, et al. Neuropsychiatric features of a cohort of patients with systemic lupus erythematosus. *ISRN Rheumatol.* 2012;989218. doi:10.5402/2012/989218

[12] Olah C, Schwartz N, Denton C, et al. Cognitive dysfunction in autoimmune rheumatic diseases. *Arthritis Res Ther.* 2020;22:78. doi:10.1186/s13075-020-02180-5

[13] Moore E, Huang MW, Putterman C. Advances in the diagnosis, pathogenesis and treatment of neuropsychiatric systemic lupus erythematosus. *Rheumatology.* 2020;32(2). doi:10.1097/BOR.0000000000000682

[14] Hanly J, Kozora E, Beyea S, Birnbaum J. Review: Nervous system disease in systemic lupus erythematosus: current status and future directions. *Arthritis Rheumatol.* 2019;71(1):33-42. doi:10.1002/art.40591

[15] Leslie B, Crowe SF. Cognitive functioning in systemic lupus erythematosus: a meta-analysis. *Lupus.* 2018;27(6):920-929. doi:10.1177/0961203317751859

[16] Zabala A, Salqueiro M, Saez-Atxukarro O, et al. Cognitive impairment in patients with neuropsychiatric and non-neuropsychiatric systemic lupus erythematosus: a systematic review and meta-analysis. *J Int Neuropsychol Soc.* 2018;24:629-639. doi:10.1017/S1355617718000073

[17] Hay EM, Black D, Huddy A, et al. Psychiatric disorder and cognitive impairment in systemic lupus erythematosus. *Arthritis Rheum.* 1992. doi:10.1002/art.1780350409

[18] Hussein H, Daker L, Fouad N, Elamir A, Mohamed S. Does vitamin D deficiency contribute to cognitive dysfunction in patients with systemic lupus erythematosus? *Innov Clin Neurosci.* 2018;15(9-10):25-29. PMID: 30588363

[19] Faria R, Goncalves J, Dias R. Neuropsychiatric systemic lupus erythematosus involvement: Towards a tailored approach to our patients? *Rambam Maimonides Med J.* 2017;8(1):e0001. doi:10.5041/RMMJ.10276

[20] The American College of Rheumatology (ACR), Ad Hoc Committee on Neuropsychiatric Lupus Nomenclature. The American College of Rheumatology nomenclature and case definitions for neuropsychiatric lupus syndromes. *Arthritis Rheum.* 1999;42(4):599-608. doi:10.1002/1529-0131(199904)42:4<599::AID-ANR2>3.0.CO;2-F

[21] Kakati S, Barman B, Ahmed S, Hussain M. Neurological manifestations in systemic lupus erythematosus: A single center study from North East India. *J Clin Diagn Res.* 2017;11(1):OC05-OC09. doi:10.7860/JCDR/2017/23773.9280

[22] Ferreira de Oliveira F, Cardoso T, Sampaio-Barros P, Damasceno B. Normal pressure hydrocephalus in the spectrum of neurological complications of systemic lupus erythematosus. *Neurol Sci.* 2013;34:1009-1013. doi:10.1007/s10072-012-1161-3

[23] Sanna G, Bertolaccini M, Khamashta M. Neuropsychiatric involvement in systemic lupus erythematosus: current therapeutic approach. *Curr Pharm Des.* 2008;14(13):1261-1269. doi:10.2174/138161208799316401

[24] Ainiala H, Loukkola J, Peltola J, Korpela M, Hietaharju A. The prevalence of neuropsychiatric syndromes in systemic lupus erythematosus. *Neurology.* 2001;57(3):496-500. doi:10.1212/wnl.57.3.496

[25] Wallace DJ, Hahn BH. *Dubois' Lupus Erythematosus.* 7th ed. Philadelphia, PA: Lippincott Williams & Wilkins; 2007. ISBN: 978-0-78-179394-0. Available at: https://www.wolterskluwer.com/en/solutions/ovid/dubois-lupus-erythematosus-3484

[26] Magro-Checa C, Zirkzee EJ, Huizinga TW, Steuo-Beekman GM. Management of neuropsychiatric systemic lupus erythematosus: Current approaches and future perspectives. *Drugs.* 2016;76:459-483. doi:10.1007/s40265-015-0534-3

[27] Hanly J, Urowitz M, Su L, et al. Autoantibodies as biomarkers for the prediction of neuropsychiatric events in systemic lupus erythematosus. *Ann Rheum Dis.* 2011;70(10):1726-1732. doi:10.1136/ard.2010.148502

[28] Central nervous system lupus: a clinical approach to therapy. *Lupus.* 2003;12(12):935-942. doi:10.1191/0961203303lu505oa

[29] Loukkola J, Laine M, Ainiala H, et al. Cognitive impairment in systemic lupus erythematosus and neuropsychiatric systemic lupus erythematosus: A population-based neuropsychological study. *J Clin Exp Neuropsychol.* 2003;25(1):145-151. doi:10.1076/jcen.25.1.145.13621

[30] Duarte-Garcia A, Romero-Diaz J, Juarez S, et al. Disease activity, autoantibodies, and inflammatory molecules in serum and cerebrospinal fluid of patients with systemic lupus erythematosus and cognitive dysfunction. *PLoS One.* 2018;13(5):e0196487. doi:10.1371/journal.pone.0196487

[31] Lefevre G, Zephir H, Warembourg F, et al. Neuropsychiatric systemic lupus erythematosus. Case definitions and diagnosis and treatment of central nervous system and psychiatric manifestations of systemic lupus erythematosus. *La Rev Med Intern.* 2012;33(9):491-502. doi:10.1016/j.revmed.2012.03.356

[32] Thurman J, Serkova N. Non-invasive imaging to monitor lupus nephritis and neuropsychiatric systemic lupus erythematosus. *F1000Research.* 2015;4:153. doi:10.12688/f1000research.6587.2

[33] Asano T, Ito H, Kariya Y, et al. Evaluation of blood-brain barrier function by quotient alpha2 macroglobulin and its relationship with interleukin-6 and complement component 3 levels in neuropsychiatric systemic lupus erythematosus. *PLoS One.* 2017;12(10):e0186414. doi:10.1371/journal.pone.0186414

[34] Zabala A, Salqueiro M, Saez-Atxukarro O, et al. Cognitive impairment in patients with neuropsychiatric and non-neuropsychiatric systemic lupus erythematosus: a systematic review and meta-analysis. *J Int Neuropsychol Soc.* 2018;24:629-639. doi:10.1017/S1355617718000073

[35] Cannerfelt B, Nystedt J, Jonsen A, et al. White matter lesions and brain atrophy in systemic lupus erythematosus patients: correlation to cognitive dysfunction in a cohort of systemic lupus erythematosus patients using different definition models for neuropsychiatric systemic lupus erythematosus. *Lupus*. 2018;27:1140-1149. doi:10.1177/0961203318763533

[36] Zhang Z, Wang Y, Shen Z, et al. The neurochemical and microstructural changes in the brain of systemic lupus erythematosus patients: A multimodal MRI study. *Sci Rep*. 2016;6:19026. doi:10.1038/srep19026

[37] Özakbaş S. Cognitive Impairment in Multiple Sclerosis: Historical Aspects, Current Status, and Beyond. *Noro Psikiyatr Ars*. 2015;52(Suppl 1):S12-S15. doi:10.5152/npa.2015.12610

[38] Sarbu N, Bargallo N, Cervera R. Advanced and conventional magnetic resonance imaging in neuropsychiatric lupus. *F1000Res*. 2015;4:162. doi:10.12688/f1000research.6522.2

[39] Qian X, Ji F, Ng KK, et al. Brain white matter extracellular free-water increases are related to reduced neurocognitive function in systemic lupus erythematosus. *Rheumatology (Oxford)*. 2021;Jun 22:511. doi:10.1093/rheumatology/keab511

[40] Nystedt J, Nilsson M, Jonsen A, et al. Altered white matter microstructure in lupus patients: a diffusion tensor imaging study. *Arthritis Res Ther*. 2018;20:21. doi:10.1186/s13075-018-1516-0

[41] Moraes-Fontes M, Lucio I, Santos C, et al. Neuropsychiatric features of a cohort of patients with systemic lupus erythematosus. *ISRN Rheumatol*. 2012;989218. doi:10.5402/2012/989218

[42] Rayes HA, Tani C, Kwan A, et al. What is the prevalence of cognitive impairment in lupus and which instruments are used to measure it? A systematic review and meta-analysis. *Semin Arthritis Rheum*. 2018;48(2):240-255. doi:10.1016/j.semarthrit.2018.02.007

[43] Kozora E, Ulug A, Erkan D, et al. Functional magnetic resonance imaging of working memory and executive dysfunction in systemic

lupus erythematosus and antiphospholipid antibody-positive patients. *Arthritis Care Res.* 2016;68:1655-1663. doi:10.1002/acr.22873

[44] Caceres V. Pinpoint cognitive dysfunction in patients with lupus. *The Rheumatologist*. Accessed April 2025. Available from: https://www.the-rheumatologist.org/article/pinpoint-cognitive-dysfunction-in-patients-with-lupus/

[45] Barraclough M, McKie S, Parker B, et al. Altered cognitive function in systemic lupus erythematosus and associations with inflammation and functional and structural brain changes. *Ann Rheum Dis.* 2019;78:934-940. doi:10.1136/annrheumdis-2018-214677

[46] El-Shafey AM, Abd-El-Geleel SM, Soliman ES. Cognitive impairment in non-neuropsychiatric systemic lupus erythematosus. *Egypt Rheumatol.* 2012;34:67-73. doi:10.1016/j.ejr.2012.02.002

[47] Muscal ET, Bloom DR, Hunter JV, Myones BL. Neurocognitive deficits and neuroimaging abnormalities are prevalent in children with lupus: Clinical and research experiences at a U.S. pediatric institution. *Lupus.* 2010;19:2679-2680. doi:10.1177/0961203309352092

[48] Butt B, Farman S, Khan S, Saeed MA, Ahmad NM. Cognitive dysfunction in patients with systemic lupus erythematosus. *Pak J Med Sci.* 2017;33(1):59-64. doi:10.12669/pjms.331.11947

[49] Olah C, Schwartz N, Denton C, Kardos Z, Putterman C, Szekanecz Z. Cognitive dysfunction in autoimmune rheumatic diseases. *Arthritis Res Ther.* 2020;22:78. doi:10.1186/s13075-020-02180-5

[50] Sahebari M, Rezaieyazdi Z, Khodashahi M, Abbasi B, Ayatollahi F. Brain single photon emission computed tomography scan (SPECT) and functional MRI in systemic lupus erythematosus patients with cognitive dysfunction: A systematic review. *Asia Oceania J Nucl Med Biol.* 2018;6(2):97-107. doi:10.22038/aojnmb.2018.26381.1184

[51] Tomietto P, Annese V, D'Agostini S, Venturini P, La Torre G, De Vita S. General and specific factors associated with severity of cognitive

impairment in systemic lupus erythematosus. *Arthritis Rheum.* 2007;57:1461-72. doi:10.1002/art.23098

[52] Liu S, Cheng Y, Zhao Y, et al. Clinical factors associated with brain volume reduction in systemic lupus erythematosus patients without major neuropsychiatric manifestations. *Front Psychiatry.* 2018;9(8)

[53] Leslie B, Crowe SF. Cognitive functioning in systemic lupus erythematosus: a meta-analysis. *Lupus.* 2018;27(6):920-29. doi:10.1177/0961203317751859

[54] Bendorius M, Po Chrystelle, Jeltsch-David H. From systemic inflammation to neuroinflammation: The case of neurolupus. *Int J Mol Sci.* 2018;19(11):3588. doi:10.3390/ijms19113588

[55] Mani A, Shenavandeh S, Sepehrtaj SS, Javadpour. Memory and learning functions in patients with systemic lupus erythematosus: A neuropsychological case-control study. *Egypt Rheumatol.* 2015;37(4):S13-S17. doi:10.1016/j.ejr.2015.02.004

[56] Geddes MR, O'Connell ME, Fisk JD, et al. Remote cognitive and behavioral assessment: Report of the Alzheimer Society of Canada Task Force on dementia care best practices for COVID-19. *Alzheimers Dement.* 2020;12(1). doi:10.1002/dad2.12111

[57] Petersen RC, Lopez O, Armstrong MJ, et al. Practice guideline update: mild cognitive impairment—report of the Guideline Development, Dissemination, and Implementation Subcommittee of the American Academy of Neurology. *Neurology.* 2018;90(3):126–135. doi:10.1212/WNL.0000000000004826

Chapter 5

[1] Barraclough M, McKie S, Parker B, Elliott R, Bruce IN. The effects of disease activity on neuronal and behavioural cognitive processes in systemic lupus erythematosus. *Lara D Veeken.* 2021;61(1):195-204. doi:10.1093/rheumatology/keab256

[2] Kello N, Anderson E, Diamond B. Cognitive dysfunction in systemic lupus erythematosus: a case for initiating trials. *Arthritis & Rheumatology*. 2019;71(9):1413-1425. doi:10.1002/art.40933

[3] Khoury LE, Zarfeshani A, Diamond B. Using the mouse to model human diseases: cognitive impairment in systemic lupus erythematosus. *The Journal of Rheumatology*. 2020;47(7):1145-1149. doi:10.3899/jrheum.200410

[4] Lampner C. Managing cognitive dysfunction in systemic lupus erythematosus. *Rheumatology Advisor*. Published 2018. Accessed April 14, 2025. https://www.rheumatologyadvisor.com/features/managing-cognitive-dysfunction-in-systemic-lupus-erythematosus/

[5] Wallace DJ, Hahn BH, eds. *Dubois' Lupus Erythematosus*. 7th ed. Philadelphia, PA: Lippincott Williams & Wilkins; 2007.

[6] Rayes HA, Tani C, Kwan A, et al. What is the prevalence of cognitive impairment in lupus and which instruments are used to measure it? A systematic review and meta-analysis. *Semin Arthritis Rheum*. 2018;48(2):240-255. doi:10.1016/j.semarthrit.2018.02.007

[7] Shucard JL, Hamlin AS, Shucard DW. The relationship between processing speed and working memory demand in systemic lupus erythematosus: evidence from a visual N-back test. *Neuropsychology*. 2011;25(1):45-52. doi:10.1037/a0021218

[8] Benigas J, Benigas JE, Brush J, Elliot G. *Spaced retrieval step by step: An Evidence-Based Mmory Intervention*.; 2016.

[9] Bogaczewicz J, Sysa-Jedrzejowska A, Arkuszewska C, et al. Vitamin D status in systemic lupus erythematosus patients and its association with selected clinical and laboratory parameters. *Lupus*. 2012;21(5):477-484. doi:10.1177/0961203311427549

[10] Conti F, Alessandri C, Perricone C, et al. Neurocognitive dysfunction in systemic lupus erythematosus: association with antiphospholipid antibodies, disease activity, and chronic damage. *PLoS One*. 2012;7(3):e33824. doi:10.1371/journal.pone.0033824

[11] Plantinga L, Tift B, Dunlop-Thomas C, et al. Geriatric assessment of physical and cognitive functioning in a diverse cohort of systemic lupus erythematosus patients: A pilot study. *Arthritis Care Res (Hoboken)*. 2018;70(10):1469-1477. doi:10.1002/acr.23507

[12] Nishimura K, Omori M, Katsumata Y, et al. Neurocognitive impairment in corticosteroid-naïve patients with active systemic lupus erythematosus: a prospective study. *J Rheumatol*. 2015;42:441-448. doi:10.3899/jrheum.140659

[13] Yue R, Gurung I, Long XX, Xian JY, Peng XB. Prevalence, involved domains, and predictor of cognitive dysfunction in systemic lupus erythematosus. *Lupus*. 2020;29(13):1743–1751. doi:10.1177/0961203320958061

[14] Meara A, Davidson N, Steigelman H, et al. Screening for cognitive impairment in SLE using the Self-Administered Gerocognitive Exam. *Lupus*. 2018;27(8):1363-1367. doi:10.1177/0961203318759429

[15] El-Shafey AM, Abd-El-Geleel SM, Soliman ES. Cognitive impairment in non-neuropsychiatric systemic lupus erythematosus. *Egypt Rheumatol*. 2012;34:67–73. doi:10.1016/j.ejr.2012.02.002

[16] Anderson EW, Fishbein J, Hong J, et al. Quinolinic acid, a kynurenine/tryptophan pathway metabolite, associates with impaired cognitive test performance in systemic lupus erythematosus. *Lupus Sci Med*. 2021;8:e000559. doi:10.1136/lupus-2021-000559

[17] Morrison E, Carpenter S, Shaw E, Doucette S, Hanly JG. Neuropsychiatric systemic lupus erythematosus: Association with global disease activity. *Lupus*. 2014;23:370–377. doi:10.1177/0961203314520843

[18] Veeranki SP, Downer B, Jupiter D, Wong R. Arthritis and risk of cognitive and functional impairment in older Mexican adults. *J Aging Health*. 2017;29(3):454–473. doi:10.1177/0898264316636838

[19] Gerosa M, Poletti B, Pregnolato F, et al. Antiglutamate receptor antibodies and cognitive impairment in primary antiphospholipid

syndrome and systemic lupus erythematosus. *Front Immunol.* 2016;7:5. doi:10.3389/fimmu.2016.00005

[20] Harrison M. Thinking, memory, and lupus: What we know and what we can do. *Hospital for Special Surgery.* Published 2006. Available at: https://www.hss.edu/conditions_thinking-memory-lupus.asp

[21] Benedict R, Shucard J, Zivadinov R, Shucard D. Neuropsychological impairment in systemic lupus erythematosus: A comparison with multiple sclerosis. *Neuropsychol Rev.* 2008;18(2):149–166. doi:10.1007/s11065-008-9061-2

[22] Chandler MJ, Parks AC, Marsiske M, Rotblatt LJ, Smith GE. Everyday impact of cognitive interventions in mild cognitive impairment: A systematic review and meta-analysis. *Neuropsychol Rev.* 2017;26(3):225–251. doi:10.1007/s11065-016-9330-4

[23] Bredesen DE. Reversal of cognitive decline: A novel therapeutic program. *Aging.* 2014;9:707–717. doi:10.18632/aging.100690

[24] Bredesen DE. *The End of Alzheimer's Program: The First Protocol to Enhance Cognition and Reverse Decline at Any Age.* New York: Penguin Random House; 2020. Available from: https://a.co/d/5LwgXlt

Chapter 6

[1] El-Shafey AM, Abd-El-Geleel SM, Soliman ES. Cognitive impairment in non-neuropsychiatric systemic lupus erythematosus. *Egypt Rheumatol.* 2012;34:67-73. doi:10.1016/j.ejr.2012.02.002

[2] Mackay M. Lupus brain fog: A biologic perspective on cognitive impairment, depression, and fatigue in systemic lupus erythematosus. *Immunol Res.* 2015;63:26-37. doi:10.1007/s12026-015-8716-3

[3] Wiseman SJ, Bastin ME, Amft EN, Belch JF, Ralston SH, Wardlaw JM. Cognitive function, disease burden and the structural connectome in systemic lupus erythematosus. *Lupus.* 2018;27(8):1329-1337. doi:10.1177/0961203318772666

[4] Zhang Y, Han H, Chu L. Neuropsychiatric lupus erythematosus: future directions and challenges; a systematic review and survey. *Clinics (Sao Paulo)*. 2020;75:e1515. doi:10.6061/clinics/2020/e1515

[5] Jung RE, Chavez RS, Flores RA, et al. White matter correlates of neuropsychological dysfunction in systemic lupus erythematosus. *PLoS One*. 2012;7(2):e28373. doi:10.1371/journal.pone.0028373

[6] Sarbu N, Alobeidi F, Toledano P, et al. Brain abnormalities in newly diagnosed neuropsychiatric lupus: systematic MRI approach and correlation with clinical and laboratory data in a large multi-center cohort. *Autoimmun Rev*. 2015;14(2):153-159. doi:10.1016/j.autrev.2014.11.001

[7] Cannerfelt B, Nystedt J, Jonsen A, et al. White matter lesions and brain atrophy in systemic lupus erythematosus patients: correlation to cognitive dysfunction in a cohort of systemic lupus erythematosus patients using different definition models for neuropsychiatric systemic lupus erythematosus. *Lupus*. 2018;27(7):1140-1149. doi:10.1177/0961203318763533

[8] Wiseman SJ, Bastin ME, Hamilton IF, et al. Fatigue and cognitive function in systemic lupus erythematosus: associations with white matter microstructural damage. A diffusion tensor MRI study and meta-analysis. *Lupus*. 2017;26(6):588-597. doi:10.1177/0961203316668417

[9] Costallat BL, Ferreira DM, Lapa AT, et al. Brain diffusion tensor MRI in systemic lupus erythematosus: a systematic review. *Autoimmun Rev*. 2018;17(1):36-43. doi:10.1016/j.autrev.2017.11.008

[10] Luyendijk J, Steens SC, Ouwendijk WJ, et al. Neuropsychiatric systemic lupus erythematosus: lessons learned from magnetic resonance imaging. *Arthritis Rheum*. 2011;63(3):722-732. doi:10.1002/art.30157

[11] Benedict R, Shucard J, Zivadinov R, Shucard D. Neuropsychological impairment in systemic lupus erythematosus: a comparison with multiple sclerosis. *Neuropsychol Rev*. 2008;18(2):149-166. doi:10.1007/s11065-008-9061-2

[12] Kivity S, Agmon-Levin N, Zandman-Goddard G, Chapman J, Shoenfeld Y. Neuropsychiatric lupus: a mosaic of clinical presentations. *BMC Med*. 2015;13:43. doi:10.1186/s12916-015-0269-8

[13] Nystedt J, Nilsson M, Jonsen A, et al. Altered white matter microstructure in lupus patients: a diffusion tensor imaging study. *Arthritis Res Ther*. 2018;20(1):21. doi:10.1186/s13075-018-1516-0

[14] Magro-Checa C, Kumar S, Ramiro S, et al. Are serum autoantibodies associated with brain changes in systemic lupus erythematosus? MRI data from the Leiden NP-SLE cohort. *Lupus*. 2019;28(1):94-103. doi:10.1177/0961203318816819

[15] Appenzeller S, Rondina JM, Li LM, et al. Cerebral and corpus callosum atrophy in systemic lupus erythematosus. *Arthritis Rheum*. 2005;52(9):2783-2789. doi:10.1002/art.21271

[16] Ni S, An N, Li C, et al. Altered structural and functional homotopic connectivity associated with cognitive changes in SLE. *Lupus Sci Med*. 2024;11(2):e001307. doi:10.1136/lupus-2024-001307

[17] Shucard JL, Hamlin AS, Shucard DW. The relationship between processing speed and working memory demand in systemic lupus erythematosus: evidence from a visual N-back test. *Neuropsychology*. 2011;25(1):45-52. doi:10.1037/a0021218

[18] Ploran E, Tang C, Mackay M, et al. Assessing cognitive impairment in SLE: examining relationships between resting glucose metabolism and anti-NMDAR antibodies with navigational performance. *Lupus Sci Med*. 2019;6:1-10. doi:10.1136/lupus-2019-000327

[19] Cao ZY, Wang N, Jia JT, et al. Abnormal topological organization in systemic lupus erythematosus: a resting-state functional magnetic resonance imaging analysis. *Brain Imaging Behav*. 2020. doi:10.1007/s11682-019-00228-y

[20] Wu B, Ma Y, Xie L, et al. Impaired decision-making and functional neuronal network activity in systemic lupus erythematosus. *J Magn Reson Imaging*. 2018;48(6):1508-1517. doi:10.1002/jmri.26006

[21] Diamond B, Volpe BT. A model for lupus brain disease. *Immunol Rev.* 2012;248(1):56-67. doi:10.1111/j.1600-065X.2012.01137.x

[22] Wang Y, Coughlin JM, Ma S, et al. Neuroimaging of translocator protein in patients with systemic lupus erythematosus: a pilot study using CDPA-713 positron emission tomography. *Lupus.* 2018;26(2):170-178. doi:10.1177/0961203316657432

[23] Zhang Z, Wang Y, Shen Z, et al. The neurochemical and microstructural changes in the brain of systemic lupus erythematosus patients: a multimodal MRI study. *Sci Rep.* 2016;6:19026. doi:10.1038/srep19026

[24] Kapadia M, Bijjelic D, Zhao H, et al. Effects of sustained i.c.v. infusion of lupus CSF and autoantibodies on behavioral phenotype and neuronal calcium signaling. *Acta Neuropathol Commun.* 2017;5. doi:10.1186/s40478-017-0473-1

[25] Kozora E, Ulug AM, Erkan D, et al. Functional magnetic resonance imaging of working memory and executive dysfunction in systemic lupus erythematosus and antiphospholipid antibody-positive patients. *Arthritis Care Res (Hoboken).* 2016;68(11):1655–1663. doi:10.1002/acr.22873

[26] Barraclough M, Elliott R, McKie S, Parker B, Bruce IN. Cognitive dysfunction and functional magnetic resonance imaging in systemic lupus erythematosus. *Lupus.* 2015;24(12):1239-1247. doi:10.1177/0961203315593 81

[27] DiFrancesco MW, Gitelman D, Klein-Gitelman MS, et al. Functional neural network activity differs with cognitive dysfunction in childhood-onset systemic lupus erythematosus. *Arthritis Res Ther.* 2013;15(2):R41. doi:10.1186/ar4197

[28] Fedorenko E, Dunan J, Kanwisher N. Broad domain generality in focal regions of frontal and parietal cortex. *Proc Natl Acad Sci U S A.* 2013;110(41):16616-16621. doi:10.1073/pnas.1315235110

[29] Ploran E, Tang C, Mackay M, et al. Assessing cognitive impairment in SLE: examining relationships between resting glucose metabolism and

anti-NMDAR antibodies with navigational performance. *Lupus Sci Med.* 2019;6:1-10. doi:10.1136/lupus-2019-000327

[30] Khoury LE, Zarfeshani A, Diamond B. Using the mouse to model human diseases: cognitive impairment in systemic lupus erythematosus. *J Rheumatol.* 2020;47(7):1145-1149. doi:10.3899/jrheum.200410

[31] Rimkus C, Steenwijk M, Barkhof F. Causes, effects, and connectivity changes in MS-related cognitive decline. *Dement Neuropsychol.* 2016;10(1):2-11. doi:10.1590/S1980-57642016DN10100002

Chapter 7

[1] Moore E, Huang MW, Putterman C. Advances in the diagnosis, pathogenesis and treatment of neuropsychiatric systemic lupus erythematosus. *Rheumatology.* 2020;32(2). doi:10.1097/BOR.0000000000000682

[2] Meara A, Davidson N, Steigelman H, Zhao S, Brock G, Jarjour WN, Rovin BH, Madhoun H, Parikh S, Hebert L, Ayoub I, Ardoin SP. Screening for cognitive impairment in SLE using the Self-Administered Gerocognitive Exam. *Lupus.* 2018;27:1363-1367. doi:10.1177/0961203318759429

[3] Lampner C. Managing cognitive dysfunction in Systemic Lupus Erythematosus. *Rheumatology Advisor.* Published 2018. Accessed at: https://www.rheumatologyadvisor.com/features/managing-cognitive-dysfunction-in-systemic-lupus-erythematosus/

[4] Popescu A, Kao AH. Neuropsychiatric systemic lupus erythematosus. *Curr Neuropharmacol.* 2011;9(3):449-457. doi:10.2174/157015911796557984

[5] Hanly J, Urowitz M, Su L, Bae S, Gordon C, Bernatsky S, Vasudevan A, Isenberg D, Rahman A, Wallace DJ, Merrill J. Autoantibodies as biomarkers for the prediction of neuropsychiatric events in systemic lupus erythematosus. *Ann Rheum Dis.* 2011;70(10):1726-1732. doi:10.1136/ard.2010.148502

[6] Wallace DJ, Hahn BH. *Dubois' Lupus Erythematosus*. 7th ed. Philadelphia, PA: Lippincott Williams & Wilkins; 2007. ISBN: 978-0-78-179394-0.

[7] Tay SH, Mak A. Anti-NR2A/B antibodies and other major molecular mechanisms in the pathogenesis of cognitive dysfunction in systemic lupus erythematosus. *Int J Mol Sci*. 2015;16(5):10281-10300. doi:10.3390/ijms160510281

[8] Sood A, Al R, Raji M. Cognitive impairment in elderly patients with rheumatic disease and the effect of disease-modifying anti-rheumatic drugs. *Clin Rheumatol*. 2021. doi:10.1007/s10067-020-05372-1

[9] Cunningham C, Hennessy E. Co-morbidity and systemic inflammation as drivers of cognitive decline: new experimental models adopting a broader paradigm in dementia research. *Alzheimers Res Ther*. 2015;7(1):33. doi:10.1186/s13195-015-0117-2

[10] Kozora E, Laudenslager M, Lemieux A, West SG. Inflammatory and hormonal measures predict neuropsychological functioning in systemic lupus erythematosus and rheumatoid arthritis patients. *J Int Neuropsychol Soc*. 2001;7(6):745-754. doi:10.1017/S1355617701766106

[11] Putterman C, Mikde E, Garcia S. Exploring the role of Lipocalin-2 in neuropsychiatric SLE pathogenesis. In: *American College of Rheumatology Convergence Meeting Abstracts*. ACR; 2020. Accessed April 14, 2025. https://acrabstracts.org

[12] Mackay M. Lupus brain fog: A biologic perspective on cognitive impairment, depression, and fatigue in systemic lupus erythematosus. *Immunol Res*. 2015;63:26-37. doi:10.1007/s12026-015-8716-3

[13] Appenzeller S, Rondina JM, Li LM, Costallat LT, Cendes F. Cerebral and corpus callosum atrophy in systemic lupus erythematosus. *Arthritis Rheum*. 2005;52(9):2783-2789. doi:10.1002/art.21271

[14] Shucard JL, Gaines JJ, Ambrus J, Shucard DW. C-reactive protein and cognitive deficits in systemic lupus erythematosus. *Cogn Behav Neurol*. 2007;20(1):31-37. doi:10.1097/WNN.0b013e31802e3b9a

[15] Wen J, Stock A, Chalmers A, Putterman C. The role of B cells and autoantibodies in neuropsychiatric lupus. *Autoimmun Rev*. 2016;15(9):890-895. doi:10.1016/j.autrev.2016.07.009

[16] Boumpas D. Management of neuropsychiatric SLE. *Lupus Sci Med*. 2022;9(Suppl 1):A8.2. https://lupus.bmj.com/content/9/Suppl_1/A8.2

[17] Schwabe L, Wolf OT, Oitzl MS. Memory formation under stress: quantity and quality. *Neurosci Biobehav Rev*. 2010;34(4):584-591. doi:10.1016/j.neubiorev.2009.11.015

[18] Aimone JB, Deng W, Gage FH. Resolving new memories: a critical look at the dentate gyrus, adult neurogenesis, and pattern separation. *Neuron*. 2011;70(4):589-596. doi:10.1016/j.neuron.2011.05.010

[19] Trysberg E, Nylen K, Rosengren LE, Tarkowski A. Neuronal and astrocytic damage in systemic lupus erythematosus patients with central nervous system involvement. *Arthritis Rheum*. 2003;48:2881-2887. doi:10.1002/art.11279

[20] Cagnoli P, Haris RE, Frechtling D, et al. Reduced insular glutamine and N-acetylaspartate in systemic lupus erythematosus. *Acad Radiol*. 2013;20(10):1286-1296. https://doi.org/10.17615/nvxn-a725

[21] Faust T, Chang E, Kowal C, et al. Neurotoxic lupus autoantibodies alter brain function through two distinct mechanisms. *Proc Natl Acad Sci U S A*. 2010;107:18569-18574. doi:10.1073/pnas.1006980107

[22] Benedict R, Shucard J, Zivadinov R, Shucard D. Neuropsychological impairment in systemic lupus erythematosus: a comparison with multiple sclerosis. *Neuropsychol Rev*. 2008;18(2):149-166. doi:10.1007/s11065-008-9061-2

[23] Opendak M, Gould E. Adult neurogenesis: a substrate for experience-dependent change. *Trends Cogn Sci*. 2015;19(3):151-161. doi:10.1016/j.tics.2015.01.001

[24] Lapa A, Postal M, Sinicato N, et al. S100B is associated with cognitive impairment in childhood-onset systemic lupus erythematosus patients. *Lupus*. 2017;26:478-483. doi:10.1177/0961203317691374

[25] Piehl N, Olst L, Ramakrishnan A, et al. Cerebrospinal fluid immune dysregulation during healthy brain aging and cognitive impairment. *Cell*. 2022;185(26). doi:10.1016/j.cell.2022.11.019

[26] Ho RC, Thiaghu C, Ong H, et al. A meta-analysis of serum and cerebrospinal fluid autoantibodies in neuropsychiatric systemic lupus erythematosus. *Autoimmun Rev*. 2016;15(2):124-138. doi:10.1016/j.autrev.2015.10.003

[27] Balajkova V, Olejarova M, Moravcova R, et al. Is serum TWEAK a useful biomarker of neuropsychiatric systemic lupus erythematosus? *Physiol Res*. 2020;69(2):339-346. doi:10.33549/physiolres.934308

[28] Kapadia M, Bijjelic D, Zhao H, Donglai M, et al. Effects of sustained i.c.v. infusion of lupus CSF and autoantibodies on behavioral phenotype and neuronal calcium signaling. *Acta Neuropathol Commun*. 2017;5. doi:10.1186/s40478-017-0473-1

[29] Allen WE, Blosser TR, Sullivan ZA, Dulac C, Zhuang X. Molecular and spatial signatures of mouse brain aging at single-cell resolution. *Cell*. 2022;186(1). doi:10.1016/j.cell.2022.12.010

[30] Duarte-Garcia A, Romero-Diaz J, Juarez S, et al. Disease activity, autoantibodies, and inflammatory molecules in serum and cerebrospinal fluid of patients with systemic lupus erythematosus and cognitive dysfunction. *PLoS One*. 2018;13(5):1-13. doi:10.1371/journal.pone.0196487

[31] Wiseman SJ, Bastin ME, Jardine CL, et al. Cerebral small vessel disease burden is increased in systemic lupus erythematosus. *Stroke*. 2016;47(11):2722-2728. doi:10.1161/STROKEAHA.116.014330

[32] Leslie B, Crowe SF. Cognitive functioning in systemic lupus erythematosus: a meta-analysis. *Lupus*. 2018;27(6):920-929. doi:10.1177/0961203317751859

[33] Jafri K, Patterson S, Lanata C. Central nervous system manifestations of systemic lupus erythematosus. *Rheum Dis Clin North Am.* 2017;43(4):531-545. doi:10.1016/j.rdc.2017.06.003

[34] Wang Y, Coughlin JM, Ma S, et al. Neuroimaging of translocator protein in patients with systemic lupus erythematosus: A pilot study using CDPA-713 Positron Emission Tomography. *Lupus.* 2018;26(2):170-178. doi:10.1177/0961203316657432

[35] Abraham P, Neel I, Bishay S, Sewell DD. Central nervous system systemic lupus erythematosus (CNS-SLE) vasculitis mimicking Lewy body dementia: A case report emphasizing the role of imaging with an analysis of 33 comparable cases from the scientific literature. *J Geriatr Psychiatry Neurol.* 2021;34(2):128-141. doi:10.1177/0891988720901788

[36] Qian X, Ji F, Ng KK, et al. Brain white matter extracellular free-water increases are related to reduced neurocognitive function in systemic lupus erythematosus. *Rheumatology (Oxford).* 2021;Jun 22:511. doi:10.1093/rheumatology/keab511

[37] Mak A, Ho R, Tng H, et al. Early cerebral volume reductions and their associations with reduced lupus disease activity in patients with newly-diagnosed systemic lupus erythematosus. *Sci Rep.* 2016;6:22231. doi:10.1038/srep22231

[38] Shastri R, Shah G, Wang P, et al. MR diffusion tractography to identify and characterize microstructural white matter tract changes in systemic lupus erythematosus patients. *Acad Radiol.* 2016;23(11):1431-1440. doi:10.1016/j.acra.2016.03.019

[39] Lee S, Wu C, Hsieh L, Cheung W, Chou MC. Efficacy of magnetic resonance diffusion tensor imaging and three-dimensional fiber tractography in the detection of clinical manifestations of central nervous system lupus. *Magn Reson Imaging.* 2014;32(5):598-603. doi:10.1016/j.mri.2014.02.005

[40] Cannerfelt B, Nystedt J, Jonsen A, et al. White matter lesions and brain atrophy in systemic lupus erythematosus patients: correlation to cognitive dysfunction in a cohort of systemic lupus erythematosus patients using different definition models for neuropsychiatric systemic

lupus erythematosus. *Lupus*. 2018;27:1140-1149. doi:10.1177/0961203318763533

[41] Song K, Han HJ, Kim S, Kwon J. Thymosin beta 4 attenuates PrP(106-126)-induced human brain endothelial cells dysfunction. *Eur J Pharmacol*. 2020;869:172891. doi:10.1016/j.ejphar.2019.172891

[42] Kamintsky L, Beyea SD, Fisk JD, et al. Blood-brain barrier leakage in systemic lupus erythematosus is associated with gray matter loss and cognitive impairment. *Ann Rheum Dis*. 2020;79(12):1580-1587. doi:10.1136/annrheumdis-2020-218004

[43] Hanly J, Robertson J, Kamintsky L, et al. Functional connectivity, enhanced blood-brain barrier leakage and cognitive impairment in systemic lupus erythematosus. Presented at: American College of Rheumatology Convergence 2021; November 3-10, 2021. Abstract 0456. https://acrabstracts.org/abstract/functional-connectivity-enhanced-blood-brain-barrier-leakage-and-cognitive-impairment-in-systemic-lupus-erythematosus/

[44] Hanly JG, Legge A, Kamintsky L, et al. Role of autoantibodies and blood-brain barrier leakage in cognitive impairment in systemic lupus erythematosus. *Lupus Sci Med*. 2022;9(1):e000668. doi:10.1136/lupus-2022-000668

[45] O'Bryan T. *You Can Fix Your Brain*. New York, NY: Penguin Random House (Rodale Books); 2018. https://a.co/d/hlbhEzA

[46] Kello N, Anderson E, Diamond B. Cognitive dysfunction in systemic lupus erythematosus: a case for initiating trials. *Arthritis Rheumatol*. 2019;71(9):1413-1425. doi:10.1002/art.40933

[47] Stock A, Gelb S, Pasternak O, Ben-Zvi A, Putterman C. The blood-brain barrier and neuropsychiatric lupus: new perspectives in light of advances in understanding the neuroimmune interface. *Autoimmun Rev*. 2017;16(6):612-619. doi:10.1016/j.autrev.2017.04.008

[48] Gelb S, Stock AD, Anzi S, Putterman C, Ben-Zvi A. Mechanisms of neuropsychiatric lupus: The relative roles of the blood-cerebrospinal fluid barrier versus blood-brain barrier. *J Autoimmun*. 2018; 91:34-44.

doi: 10.1016/j.jaut.2018.03.00110.1016/j.jaut.2018.03.001. Epub 2018 Apr 4. PMID: 29627289; PMCID: PMC5994369.

[49] Bendorius M, Po C, Jeltsch-David H. From systemic inflammation to neuroinflammation: the case of neurolupus. *Int J Mol Sci.* 2018;19(11):3588. doi:10.3390/ijms19113588

[50] Mackay M, Bussa MP, Aranow C, et al. Differences in regional brain activation patterns assessed by functional magnetic resonance imaging in patients with systemic lupus erythematosus stratified by disease duration. *Mol Med.* 2011;17(11-12):1349–1356. doi:10.2119/molmed.2011.00185

[51] Nalakonda G, Islam M, Chukwu V, Soliman A, Munim R, Abukraa I. Psycho-rheumatic integration in systemic lupus erythematosus: an insight into antibodies causing neuropsychiatric changes. *Cureus.* 2018;10(8):e3091. doi:10.7759/cureus.3091

[52] Wen J, Doerner J, Chalmers S, Stock A, Wang H. B cell and/or autoantibody deficiency do not prevent neuropsychiatric disease in murine systemic lupus erythematosus. *J Neuroinflammation.* 2016;13:1. doi:10.1186/s12974-016-0537-3

[53] Faria R, Goncalves J, Dias R. Neuropsychiatric systemic lupus erythematosus involvement: towards a tailored approach to our patients? *Rambam Maimonides Med J.* 2017;8(1):e0001. doi:10.5041/RMMJ.10276

[54] Kivity S, Agmon-Levin N, Zandman-Goddard G, Chapman J, Shoenfeld Y. Neuropsychiatric lupus: a mosaic of clinical presentations. *BMC Med.* 2015;13:43. doi:10.1186/s12916-015-0269-8

[55] Asano T, Ito H, Kariya Y, et al. Evaluation of blood-brain barrier function by quotient alpha2 macroglobulin and its relationship with interleukin-6 and complement component 3 levels in neuropsychiatric systemic lupus erythematosus. *PLoS One.* 2017;12(10):e0186414. doi:10.1371/journal.pone.0186414

[56] Kello N, Anderson E, Diamond B. Cognitive dysfunction in systemic lupus erythematosus: a case for initiating trials. *Arthritis Rheumatol*. 2019;71(9):1413-1425. doi:10.1002/art.40933

[57] Szmyrka M, Pokryszko-Dragan A, Slotwinski K, et al. Cognitive impairment, event-related potentials and immunological status in patients with systemic lupus erythematosus. *Adv Clin Exp Med*. 2018;28(2). doi:10.17219/acem/7671

[58] Wiseman SJ, Bastin ME, Hamilton IF, et al. Fatigue and cognitive function in systemic lupus erythematosus: associations with white matter microstructural damage. A diffusion tensor MRI study and meta-analysis. *Lupus*. 2017;26:588-597. doi:10.1177/0961203316668417

[59] Ceccarelli F, Perricone C, Pirone C, et al. Cognitive dysfunction improves in systemic lupus erythematosus: results of a 10-year prospective study. *PLoS One*. 2018;13(5):e0196103. doi:10.1371/journal.pone.0196103

[60] Caceres, V. Pinpoint cognitive dysfunction in patients with lupus. The Rheumatologist. Published 2012. Accessed April 15, 2025. https://www.the-rheumatologist.org/article/pinpoint-cognitive-dysfunction-in-patients-with-lupus/

[61] Govoni, M., Bortoluzzi, A., Padovan, M., Silvagni, E., Borrelli, M., Donelli, F., Ceruti, S., & Rotta, F. The diagnosis and clinical management of the neuropsychiatric manifestations of lupus. *J Autoimmun*. 2016;74:41-72. doi:10.1016/j.jaut.2016.06.013

[62] Bendorius M, Po C, Jeltsch-David H. From systemic inflammation to neuroinflammation: the case of neurolupus. *Int J Mol Sci*. 2018;19(11):3588. doi:10.3390/ijms19113588

[63] Magro-Checa C, Zirkzee EJ, Huizinga TW, Steuo-Beekman GM. Management of neuropsychiatric systemic lupus erythematosus: Current approaches and future perspectives. *Drugs*. 2016;76(5):459-483. doi:10.1007/s40265-015-0534-3

[64] Yoshimura A, Ohyagi M, Ito M. T cells in brain inflammation. *Adv Immunol*. 2022. doi:10.1016/bs.ai.2022.10.001.

[65] Fava A, Petri M. Systemic lupus erythematosus: Diagnosis and clinical management. *J Autoimmun*. 2019;96:1-13. doi:10.1016/j.jaut.2018.11.

[66] Scharer CD, Blalock EL, Tian M, et al. Epigenetic programming underpins B cell dysfunction in human SLE. *Nat Immunol*. 2019;20(8):1027-1037. doi:10.1038/s41590-019-0419-9.

[67] Olsen, N.J., & Karp, D.R. (2020). Finding lupus in the ANA haystack. British Medical Journal, 7(1). 10.1136/lupus-2020-000384

[68] Merrill JT. Research highlights from 13th international congress on lupus. *Lupus Foundation of America*. Published 2019. Accessed April 14, 2025. https://www.lupus.org/news/research-highlights-from-13th-international-congress-on-lupus

[69] Celhar T, Fairhurst A. Modelling clinical systemic lupus erythematosus: similarities, differences and success stories. *Rheumatology*. 2017;56(1):i88-i99. doi:10.1093/rheumatology/kew400

[70] Fava A, Petri M. Systemic lupus erythematosus: Diagnosis and clinical management. *J Autoimmun*. 2019;96:1-13. doi:10.1016/j.jaut.2018.11

[71] Banchereau R, Hong S, Cantarel B, et al. Personalized immunomonitoring uncovers molecular networks that stratify lupus patients. *Cell*. 2016;165(3):551-565. doi:10.1016/j.cell.2016.05.057

[72] Ploran E, Tang C, Mackay M, et al. Assessing cognitive impairment in SLE: examining relationships between resting glucose metabolism and anti-NMDAR antibodies with navigational performance. *Lupus Sci Med*. 2019;6:1-10. doi:10.1136/lupus-2019-000327

[73] Varley JA, Andersson M, Grant E, et al. Absence of neuronal autoantibodies in neuropsychiatric systemic lupus erythematosus. *Ann Neurol*. 2020;88(6):1244-1250. doi:10.1002/ana.25908

[74] Wen J, Doerner J, Chalmers S, Stock A, Wang H. B cell and/or autoantibody deficiency do not prevent neuropsychiatric disease in

murine systemic lupus erythematosus. *J Neuroinflammation*. 2016;13. doi:10.1186/s12974-016-0537-3

[75] Kalim H, Pratama M, Mahardini E, Winoto ES, Adi-Krisna P, Handono K. Accelerated immune aging was correlated with lupus-associated brain fog in reproductive-age systemic lupus erythematosus patients. *Int J Rheum Dis*. 2020;23:620-626. doi:10.1111/1756-185X.13816

[76] Kakati S, Barman B, Ahmed S, Hussain M. Neurological manifestations in systemic lupus erythematosus: A single center study from North East India. *J Clin Diagn Res*. 2017;11(1):OC05-OC09. doi:10.7860/JCDR/2017/23773.9280

[77] Tay SH, Mak A. Anti-NR2A/B antibodies and other major molecular mechanisms in the pathogenesis of cognitive dysfunction in systemic lupus erythematosus. *Int J Mol Sci*. 2015;16(5):10281-10300. doi:10.3390/ijms160510281

[78] Mike EV, Makinde HM, Der E, Stock A, Gulinello M, Gadhvi GT, Winter DR, Cuda CM, Putterman C. Neuropsychiatric systemic lupus erythematosus is dependent on Sphingosine-1-Phosphate signaling. *Front Immunol*. 2018;9:2198. doi:10.3389/fimmu.2018.02189

[79] Olah C, Schwartz N, Denton C, Kardos Z, Putterman C, Szekanecz Z. Cognitive dysfunction in autoimmune rheumatic diseases. *Arthritis Res Ther*. 2020;22:78. doi:10.1186/s13075-020-02180-5

[80] Cuffari B. What is neuropsychiatric lupus (NPSLE)? News-Medical.net. https://www.news-medical.net/health/What-is-Neuropsychiatric-Lupus-(NPSLE).aspx. Published March 2021. Accessed April 15, 2025.

[81] Barraclough M, McKie S, Parker B, et al. Altered cognitive function in systemic lupus erythematosus and associations with inflammation and functional and structural brain changes. *Ann Rheum Dis*. 2019;78:934-940. doi:10.1136/annrheumdis-2018-214677

Chapter 8

[1] Cruz DP. Clinical review: Systemic lupus erythematosus. *BMJ*. 2006;332:890-894. doi:10.1136/bmj.332.7546.890

[2] Bendorius M, Po C, Jeltsch-David H. From systemic inflammation to neuroinflammation: The case of neurolupus. *Int J Mol Sci*. 2018;19(11):3588. doi:10.3390/ijms19113588

[3] Ploran E, Tang C, Mackay M, et al. Assessing cognitive impairment in SLE: examining relationships between resting glucose metabolism and anti-NMDAR antibodies with navigational performance. *Lupus Sci Med*. 2019;6:1-10. doi:10.1136/lupus-2019-000327

[4] Thomas DE. *The Lupus Encyclopedia: A Comprehensive Guide for Patients and Families*. Baltimore, MD: Johns Hopkins Press Health; 2014. Available at: https://a.co/d/8Vp2WES.

[5] Sood A, Raji M. Cognitive impairment in elderly patients with rheumatic disease and the effect of disease-modifying anti-rheumatic drugs. *Clin Rheumatol*. 2021. doi:10.1007/s10067-020-05372-1

[6] Olah C, Schwartz N, Denton C, Kardos Z, Putterman C, Szekanecz Z. Cognitive dysfunction in autoimmune rheumatic diseases. *Arthritis Res Ther*. 2020;22:78. doi:10.1186/s13075-020-02180-5

[7] Barraclough M, McKie S, Parker B, et al. Altered cognitive function in systemic lupus erythematosus and associations with inflammation and functional and structural brain changes. *Ann Rheum Dis*. 2019;78:934-940. doi:10.1136/annrheumdis-2018-214677

[8] Leffers HC, Lange T, Collins C, Ulff-Møller CJ, Jacobsen S. The study of interactions between genome and exposome in the development of systemic lupus erythematosus. *Autoimmun Rev*. 2019;18(4):382-392. doi:10.1016/j.autrev.2018.11.005

[9] Felten R, Sagez F, Gavand P, et al. 10 most important contemporary challenges in the management of SLE. *Lupus Sci Med*. 2019;6:e000303. doi:10.1136/lupus-2018-000303

[10] Tay SH, Mak A. Anti-NR2A/B antibodies and other major molecular mechanisms in the pathogenesis of cognitive dysfunction in systemic lupus erythematosus. *Int J Mol Sci*. 2015;16(5):10281-10300. doi:10.3390/ijms160510281

[11] Magro-Checa C, Seup-Beekman G, Huizinga T, van Buchem MA, Ronen I. Laboratory and neuroimaging biomarkers in neuropsychiatric systemic lupus erythematosus: Where do we stand, where to go? *Front Med (Lausanne)*. 2018;5:340. doi:10.3389/fmed.2018.00340

[12] Cao ZY, Wang N, Jia JT, et al. Abnormal topological organization in systemic lupus erythematosus: a resting-state functional magnetic resonance imaging analysis. *Brain Imaging Behav*. 2021;15(1):14-24. doi:10.1007/s11682-019-00228-y. PMID: 31903526.

[13] Fanouriakis A, Bertsias G. Changing paradigms in the treatment of systemic lupus erythematosus. *BMJ Open*. 2018;6(1):e000310. doi:10.1136/lupus-2018-000310.

[14] Zabala A, Salqueiro M, Saez-Atxukarro O, et al. Cognitive impairment in patients with neuropsychiatric and non-neuropsychiatric systemic lupus erythematosus: a systematic review and meta-analysis. *J Int Neuropsychol Soc*. 2018;24(7):629-639. doi:10.1017/S1355617718000073

[15] The American College of Rheumatology (ACR), Ad Hoc Committee on Neuropsychiatric Lupus Nomenclature. The American College of Rheumatology nomenclature and case definitions for neuropsychiatric lupus syndromes. *Arthritis Rheum*. 1999;42(4):599-608. doi:10.1002/1529-0131(199904)42:4<599::AID-ANR2>3.0.CO;2-F

[16] Kivity S, Agmon-Levin N, Zandman-Goddard G, Chapman J, Shoenfeld Y. Neuropsychiatric lupus: a mosaic of clinical presentations. *BMC Med*. 2015;13:43. doi:10.1186/s12916-015-0269-8

[17] Yuen K, Bingham K, Tayer-Shifman O, Touma Z. Measures of cognition in rheumatic disease. *Arthritis Care Res (Hoboken).* 2020;72(S10):660-675. doi:10.1002/acr.24364

[18] Hanly J, Omisade A, Su L, Farewell V, Fisk JD. Cognitive function in systemic lupus erythematosus, rheumatoid arthritis, and multiple sclerosis assessed by computerized neuropsychological tests. *Arthritis Rheumatol.* 2010;62(5):1478-1486. doi:10.1002/art.27404

[19] Cannerfelt B, Nystedt J, Jonsen A, et al. White matter lesions and brain atrophy in systemic lupus erythematosus patients: correlation to cognitive dysfunction in a cohort of systemic lupus erythematosus patients using different definition models for neuropsychiatric systemic lupus erythematosus. *Lupus.* 2018;27(9):1140-1149. doi:10.1177/0961203318763533

[20] Bogaczewicz J, Sysa-Jedrzejowska A, Arkuszewska C, et al. Vitamin D status in systemic lupus erythematosus patients and its association with selected clinical and laboratory parameters. *Lupus.* 2012;21(5):477-484. doi:10.1177/0961203311427549

[21] Rayes HA, Tani C, Kwan A, et al. What is the prevalence of cognitive impairment in lupus and which instruments are used to measure it? A systematic review and meta-analysis. *Semin Arthritis Rheum.* 2018;48(2):240-255. doi:10.1016/j.semarthrit.2018.02.007

[22] Lampner C. Managing cognitive dysfunction in systemic lupus erythematosus. *Rheumatology Advisor.* Published 2018. Accessed April 16, 2025. https://www.rheumatologyadvisor.com/features/managing-cognitive-dysfunction-in-systemic-lupus-erythematosus

[23] Özakbaş S. Cognitive Impairment in Multiple Sclerosis: Historical Aspects, Current Status, and Beyond. *Noro Psikiyatr Ars.* 2015;52(Suppl 1):S12-S15. doi:10.5152/npa.2015.12610

[24] Meara A, Davidson N, Steigelman H, et al. Screening for cognitive impairment in SLE using the Self-Administered Gerocognitive Exam. *Lupus.* 2018;27(12):1363-1367. doi:10.1177/0961203318759429

[25] Tayer-Shifman OE, Green R, Beaton DE, et al. Validity evidence for the use of automated neuropsychologic assessment metrics as a screening tool for cognitive impairment in systemic lupus erythematosus. *Arthritis Care Res (Hoboken)*. 2020;72(12):1809-1819. doi:10.1002/acr.24096

[26] Tao R, Dhanasekaran P, Tay S, Mak A. Mathematical processing is affected by daily but not cumulative glucocorticoid dose in patients with systemic lupus erythematosus. *Rheumatology*. 2020;59(12):2534-2543. doi:10.1093/rheumatology/keaa002

[27] Paez-Venegas N, Jordan-Estrada B, Chavarria-Avila E, et al. The Montreal Cognitive Assessment Test: A useful tool in screening of cognitive impairment in patients with systemic lupus erythematosus. *J Clin Rheumatol*. Published 2018. Accessed April 16, 2025. https://acrabstracts.org/abstract/the-montreal-cognitive-assessment-test-a-useful-tool-in-screening-of-cognitive-impairment-in-patients-with-systemic-lupus-erythematosusshort-title-cognitive-impairment-in-sle/

[28] Cummings J, Aisen P, Apostolova LG, et al. Aducanumab: Appropriate use recommendations. *J Prev Alzheimers Dis*. 2021;8(4):398-410. doi:10.14283/jpad.2021.41

[29] Ozer S, Young J, Champ C, Burke M. A systematic review of the diagnostic test accuracy of brief cognitive tests to detect amnestic mild cognitive impairment. *Int J Geriatr Psychiatry*. 2016;31(11):1139-1150. doi:10.1002/gps.4444

[30] Zhang Z, Wang Y, Shen Z, et al. The neurochemical and microstructural changes in the brain of systemic lupus erythematosus patients: a multimodal MRI study. *Sci Rep*. 2016;6:19026. doi:10.1038/srep19026

[31] El-Shafey AM, Abd-El-Geleel SM, Soliman ES. Cognitive impairment in non-neuropsychiatric systemic lupus erythematosus. *Egypt Rheumatol*. 2012;34(2):67-73. doi:10.1016/j.ejr.2012.02.002

[32] Chalhoub NE, Luggen ME. Screening for cognitive dysfunction in systemic lupus erythematosus: the Montreal Cognitive Assessment

Questionnaire and the Informant Questionnaire on Cognitive Decline in the Elderly. *Lupus.* 2019;28(1):51-58. doi:10.1177/0961203318815299

[33] Moore E, Huang MW, Putterman C. Advances in the diagnosis, pathogenesis and treatment of neuropsychiatric systemic lupus erythematosus. *Rheumatology (Oxford).* 2020;32(2). doi:10.1097/BOR.0000000000000682

[34] Nantes S, Su J, Dhaliwal A, Colosimo K, Touma Z. Performance of screening tests for cognitive impairment in systemic lupus erythematosus. *J Rheumatol.* 2017;44(11):1583-1589. doi:10.3899/jrheum.161125

[35] Kanapathy A, Jaafar NR, Shaharir SS, et al. Prevalence of cognitive impairment using the Montreal Cognitive Assessment questionnaire among patients with systemic lupus erythematosus: a cross-sectional study at two tertiary centres in Malaysia. *Lupus.* 2019;28(7):854-861. doi:10.1177/0961203319852153

[36] Geddes MR, O'Connell ME, Fisk JD, et al. Remote cognitive and behavioral assessment: report of the Alzheimer Society of Canada Task Force on dementia care best practices for COVID-19. *Alzheimers Dement (Amst).* 2020;12(1):e12111. doi:10.1002/dad2.12111

[37] Raghunath S, Glikmann-Johnston Y, Morand E, Stout JC, Hoi A. Evaluation of the Montreal Cognitive Assessment as a screening tool for cognitive dysfunction in SLE. *Lupus Sci Med.* 2021;8:e000580. doi:10.1136/lupus-2021-000580

[38] Ploran E, Tang C, Mackay M, et al. Assessing cognitive impairment in SLE: examining relationships between resting glucose metabolism and anti-NMDAR antibodies with navigational performance. *Lupus Sci Med.* 2019;6:1-10. doi:10.1136/lupus-2019-000327

[39] Kozora E, Arciniegas DB, Duggan E, et al. White matter abnormalities and working memory impairment in systemic lupus erythematosus. *Cogn Behav Neurol.* 2013;26(2):63-72. doi:10.1097/WNN.0b013e31829d5c74

[40] Noguchi-Shinohara M, Domoto C, Yoshida T, et al. A new computerized assessment battery for cognition (C-ABC) to detect mild

cognitive impairment and dementia around 5 min. *PLoS One.* 2020;15(12):e0243469. doi:10.1371/journal.pone.0243469

[41] Scharre DQ, Chang S, Nagaraja HN, et al. Self-Administered Gerocognitive Examination: longitudinal cohort testing for the early detection of dementia conversion. *Alzheimers Res Ther.* 2021;13:101. doi:10.1186/s13195-021-00930-4

[42] Ozer S, Young J, Champ C, Burke M. A systematic review of the diagnostic test accuracy of brief cognitive tests to detect amnestic mild cognitive impairment. *Int J Geriatr Psychiatry.* 2016;31(1):1139-1150. doi:10.1002/gps.4444

[43] Peterson RC, Lopez O, Armstrong MJ, et al. Practice guideline update: Mild cognitive impairment. Report of the Guideline Development, Dissemination, and Implementation Subcommittee of the American Academy of Neurology. *Neurology.* 2017;doi:10.1212/WNL.0000000000004826

[44] Bogaczewicz J, Sysa-Jedrzejowska A, Arkuszewska C, et al. Vitamin D status in systemic lupus erythematosus patients and its association with selected clinical and laboratory parameters. *Lupus.* 2012;21(5):477-484. doi:10.1177/0961203311427549

[45] Julian LJ, Yazdany J, Trupin L, et al. Validity of brief screening tools for cognitive impairment in rheumatoid arthritis and systemic lupus erythematosus. *Arthritis Care Res (Hoboken).* 2012;64(3):448-454. doi:10.1002/acr.21566

[46] Plantinga L, Lim S, Bowling C, Drenkard C. Perceived stress and reported cognitive symptoms among Georgia patients with systemic lupus erythematosus. *Lupus.* 2017;26(10):1064-1071. doi:10.1177/0961203317693095

[47] Strand V, Simon LS, Meara AS, Touma Z. Measurement properties of selected patient-reported outcome measures for use in randomized controlled trials in patients with systemic lupus erythematosus: a systematic review. *Lupus Sci Med.* 2020;7(1):e000373. doi:10.1136/lupus-2019-000373

[48] Okamoto H, Kobayashi A, Yamanaka H. Cytokines and chemokines in neuropsychiatric syndromes of systemic lupus erythematosus. *J Biomed Biotechnol*. 2010;2010:268436. doi:10.1155/2010/268436

[49] Tay SH, Mak A. Diagnosing and attributing neuropsychiatric events to systemic lupus erythematosus: time to untie the Gordian knot? *Rheumatology (Oxford)*. 2017;56(1):14-23. doi:10.1093/rheumatology/kew338

[50] Bortoluzzi A, Scire C, Govoni M. Attribution of neuropsychiatric manifestations to systemic lupus erythematosus. *Front Med (Lausanne)*. 2018;5:68. doi:10.3389/fmed.2018.00068

[51] Khoury LE, Zarfeshani A, Diamond B. Using the mouse to model human diseases: cognitive impairment in systemic lupus erythematosus. *J Rheumatol*. 2020;47(7):1145-1149. doi:10.3899/jrheum.200410

[52] Barraclough M, McKie S, Parker B, et al. The effects of disease activity on neuronal and behavioural cognitive processes in systemic lupus erythematosus. *Rheumatology (Oxford)*. 2021;keab256. doi:10.1093/rheumatology/keab256

[53] Siegmann E, Muller H, Luecke C, et al. Association of depression and anxiety disorders with autoimmune thyroiditis: a systematic review and meta-analysis. *JAMA Psychiatry*. 2018;75(6):577-584. doi:10.1001/jamapsychiatry.2018.0190

[54] Mani A, Shenavandeh S, Sepehrtaj SS, Javadpour. Memory and learning functions in patients with systemic lupus erythematosus: a neuropsychological case-control study. *Egypt Rheumatol*. 2015;37(4):S13-S17. doi:10.1016/j.ejr.2015.02.004

[55] Appenzeller S, Rondina JM, Li LM, Costallat LT, Cendes F. Cerebral and corpus callosum atrophy in systemic lupus erythematosus. *Arthritis Rheum*. 2005;52(9):2783-2789. doi:10.1002/art.21271

[56] Romero-Diaz J, Isenberg D, Ramsey-Goldman R. Measures of adult systemic lupus erythematosus. *Arthritis Care Res (Hoboken)*. 2011;63(S11):S37–S49. doi:10.1002/acr.20572

[57] Plantinga L, Tift B, Dunlop-Thomas C, et al. Geriatric assessment of physical and cognitive functioning in a diverse cohort of systemic lupus erythematosus patients: a pilot study. *Arthritis Care Res (Hoboken)*. 2018;70(10):1469-1477. doi:10.1002/acr.23507

[58] Mackay M, Bussa MP, Aranow C, et al. Differences in regional brain activation patterns assessed by functional magnetic resonance imaging in patients with systemic lupus erythematosus stratified by disease duration. *Mol Med*. 2011;17(11-12):1349-1356. doi:10.2119/molmed.2011.00185

[59] Schmidt-Wilcke T, Cagnoli P, Wang P, et al. Diminished white matter integrity in patients with systemic lupus erythematosus. *Neuroimage Clin*. 2014;5:291-297. doi:10.1016/j.nicl.2014.07.001

[60] Mak A, Ho R, Tng H, et al. Early cerebral volume reductions and their associations with reduced lupus disease activity in patients with newly diagnosed systemic lupus erythematosus. *Sci Rep*. 2016;6:22231. doi:10.1038/srep22231

[61] Sahebari M, Rezaieyazdi Z, Khodashahi M, Abbasi B, Ayatollahi F. Brain single photon emission computed tomography scan (SPECT) and functional MRI in systemic lupus erythematosus patients with cognitive dysfunction: a systematic review. *Asia Ocean J Nucl Med Biol*. 2018;6(2):97–107. doi:10.22038/aojnmb.2018.26381.1184

[62] Magro-Checa C, Zirkzee EJ, Huizinga TW, Steup-Beekman GM. Management of neuropsychiatric systemic lupus erythematosus: current approaches and future perspectives. *Drugs*. 2016;76:459-483. doi:10.1007/s40265-015-0534-3

[63] Jia J, Xie J, Li H, et al. Cerebral blood flow abnormalities in neuropsychiatric systemic lupus erythematosus. *Lupus*. 2019;28(9):1128-1133. doi:10.1177/0961203319861677

[64] Castellino G, Padovan M, Bortoluzzi A, et al. Single photon emission computed tomography and magnetic resonance imaging evaluation in SLE patients with or without neuropsychiatric involvement. *Rheumatology (Oxford)*. 2008;47(3):319-323. doi:10.1093/rheumatology/kem354

[65] Waterloo K, Omdal R, Sjoholm H, et al. Neuropsychological dysfunction in systemic lupus erythematosus is not associated with changes in cerebral blood flow. *J Neurol.* 2001;248(7):595-602. doi:10.1007/s004150170138

[66] Cannerfelt B, Nystedt J, Jonsen A, et al. White matter lesions and brain atrophy in systemic lupus erythematosus patients: correlation to cognitive dysfunction in a cohort of systemic lupus erythematosus patients using different definition models for neuropsychiatric systemic lupus erythematosus. *Lupus.* 2018;27:1140-1149. doi:10.1177/0961203318763533

[67] Zaky MR, Shaat RM, El-Bassiony SR, et al. Magnetic resonance imaging brain abnormalities of neuropsychiatric systemic lupus erythematosus patients in Mansoura city: relation to disease activity. *The Egyptian Rheumatol.* 2015;37(4):S7-S11

[68] Liu S, Cheng Y, Zhao Y, et al. Clinical factors associated with brain volume reduction in systemic lupus erythematosus patients without major neuropsychiatric manifestations. *Front Psychiatry.* 2018;9(8)

[69] Shucard JL, Hamlin AS, Shucard DW. The relationship between processing speed and working memory demand in systemic lupus erythematosus: evidence from a visual N-back test. *Neuropsychology.* 2011;25(1):45-52. doi:10.1037/a0021218

[70] Toledano P, Sarbu N, Espinosa G, et al. Neuropsychiatric systemic lupus erythematosus: magnetic resonance imaging findings and correlation with clinical and immunological features. *Autoimmun Rev.* 2013;12(12):1166-1170. doi:10.1016/j.autrev.2013.07.004

[71] Benedict R, Shucard J, Zivadinov R, Shucard D. Neuropsychological impairment in systemic lupus erythematosus: a comparison with multiple sclerosis. *Neuropsychol Rev.* 2008;18(2):149–166. doi:10.1007/s11065-008-9061-2

[72] Wiseman SJ, Bastin ME, Hamilton IF, et al. Fatigue and cognitive function in systemic lupus erythematosus: associations with white matter

microstructural damage. A diffusion tensor MRI study and meta-analysis. *Lupus.* 2017;26:588-597. doi:10.1177/0961203316668417

[73] Mavrogeni S, Koutsogeorgopoulou L, Dimitroulas T, et al. Combined brain/heart magnetic resonance imaging in systemic lupus erythematosus. *Curr Cardiol Rev.* 2020;16(3):178-186. doi:10.2174/1573403X15666190801122105

[74] Sarbu N, Bargallo N, Cervera R. Advanced and conventional magnetic resonance imaging in neuropsychiatric lupus. *F1000Res.* 2015;4:162. doi:10.12688/f1000research.6522.2

[75] Szmyrka M, Pokryszko-Dragan A, Slotwinski K, et al. Cognitive impairment, event-related potentials and immunological status in patients with systemic lupus erythematosus. *Adv Clin Exp Med.* 2019;28(2):213-220. doi:10.17219/acem/76711

[76] Wang PI, Cagnoli PC, McCune WJ, et al. Perfusion-weighted MR imaging in cerebral lupus erythematosus. *Acad Radiol.* 2012;19(8):965-970. doi:10.1016/j.acra.2012.03.023

[77] Thurman J, Serkova N. Non-invasive imaging to monitor lupus nephritis and neuropsychiatric systemic lupus erythematosus. *F1000Res.* 2015;4:153. doi:10.12688/f1000research.6587.2

[78] Abraham P, Neel I, Bishay S, Sewell DD. Central nervous system systemic lupus erythematosus (CNS-SLE) vasculitis mimicking Lewy body dementia: a case report emphasizing the role of imaging with an analysis of 33 comparable cases from the scientific literature. *J Geriatr Psychiatry Neurol.* 2021;34(2):128-141. doi:10.1177/0891988720901788

[79] Sahebari M, Rezaieyazdi Z, Khodashahi M, Abbasi B, Ayatollahi F. Brain single photon emission computed tomography scan (SPECT) and functional MRI in systemic lupus erythematosus patients with cognitive dysfunction: a systematic review. *Asia Ocean J Nucl Med Biol.* 2018;6(2):97-107. doi:10.22038/aojnmb.2018.26381.1184

[80] Zardi EM, Taccone A, Marigliano B, Margiotta D, Afeltra A. Neuropsychiatric systemic lupus erythematosus: tools for the diagnosis. *Autoimmun Rev.* 2014;13(8):831-839. doi:10.1016/j.autrev.2014.04.002

[81] Piga M, Peltz MT, Montaldo C, et al. Twenty-year brain magnetic resonance imaging follow-up study in systemic lupus erythematosus: factors associated with accrual of damage and central nervous system involvement. *Autoimmun Rev.* 2015;14(6):510-516. doi:10.1016/j.autrev.2015.01.010

[82] Cannerfelt B, Nystedt J, Jonsen A, et al. White matter lesions and brain atrophy in systemic lupus erythematosus patients: correlation to cognitive dysfunction in a cohort of systemic lupus erythematosus patients using different definition models for neuropsychiatric systemic lupus erythematosus. *Lupus.* 2018;27:1140-1149. doi:10.1177/0961203318763533

[83] Govoni M, Bortoluzzi A, Padovan M, et al. The diagnosis and clinical management of the neuropsychiatric manifestations of lupus. *J Autoimmun.* 2016;74:41-72. doi:10.1016/j.jaut.2016.06.013

[84] Bendorius M, Po C, Jeltsch-David H. From systemic inflammation to neuroinflammation: the case of neurolupus. *Int J Mol Sci.* 2018;19(11):3588. doi:10.3390/ijms19113588

[85] Kozora E, Arciniegas DB, Duggan E, et al. White matter abnormalities and working memory impairment in systemic lupus erythematosus. *Cogn Behav Neurol.* 2013;26(2):63-72. doi:10.1097/WNN.0b013e31829d5c74

[86] Zardi EM, Taccone A, Marigliano B, Margiotta D, Afeltra A. Neuropsychiatric systemic lupus erythematosus: tools for the diagnosis. *Autoimmun Rev.* 2014;13(8):831–839. doi:10.1016/j.autrev.2014.04.002

Chapter 9

[1] Hanly J, Kozora E, Beyea S, Birnbaum J. Review: Nervous system disease in systemic lupus erythematosus: current status and future directions. *Arthritis Rheumatol.* 2019;71(1):33-42. doi:10.1002/art.40591

[2] Chalhoub NE, Luggen ME. Screening for cognitive dysfunction in systemic lupus erythematosus: the Montreal Cognitive Assessment Questionnaire and the Informant Questionnaire on Cognitive Decline in the Elderly. *Lupus.* 2019;28(1):51-58. doi:10.1177/0961203318815299

[3] Cannerfelt B, Nystedt J, Jonsen A, et al. White matter lesions and brain atrophy in systemic lupus erythematosus patients: correlation to cognitive dysfunction in a cohort of systemic lupus erythematosus patients using different definition models for neuropsychiatric systemic lupus erythematosus. *Lupus.* 2018;27:1140-1149. doi:10.1177/0961203318763533

[4] Nantes S, Su J, Dhaliwal A, Colosimo K, Touma Z. Performance of screening tests for cognitive impairment in systemic lupus erythematosus. *J Rheumatol.* 2017;44(11):1583-1589. doi:10.3899/jrheum.161125

[5] Leslie B, Crowe SF. Cognitive functioning in systemic lupus erythematosus: a meta-analysis. *Lupus.* 2018;27(6):920-929. doi:10.1177/0961203317751859

[6] Lampner C. Managing cognitive dysfunction in systemic lupus erythematosus. *Rheumatology Advisor.* Published 2018. Accessed April 16, 2025. https://www.rheumatologyadvisor.com/features/managing-cognitive-dysfunction-in-systemic-lupus-erythematosus/

[7] Barraclough M, Erdman L, Diaz-Martinez JP, et al. Systemic lupus erythematosus phenotypes formed from machine learning with a specific focus on cognitive impairment. *Rheumatology (Oxford).* Published online 2022. doi:10.1093/rheumatology/keac653

[8] Appenzeller S, Rondina JM, Li LM, Costallat LT, Cendes F. Cerebral and corpus callosum atrophy in systemic lupus erythematosus. *Arthritis Rheum.* 2005;52(9):2783-2789. doi:10.1002/art.21271

[9] Kivity, S., Agmon-Levin, N., Zandman-Goddard, G., Chapman, J., & Shoenfeld, Y. (2015). Neuropsychiatric lupus: a mosaic of clinical presentations. *British Medical Journal*, 13(43). DOI: 10.1186/s12916-015-0269-8

[10] Rayes HA, Tani C, Kwan A, et al. What is the prevalence of cognitive impairment in lupus and which instruments are used to measure it? A systematic review and meta-analysis. *Semin Arthritis Rheum.* 2018;48(2):240-255. doi:10.1016/j.semarthrit.2018.02.007

[11] Butt B, Farman S, Khan S, Saeed MA, Ahmad NM. Cognitive dysfunction in patients with systemic lupus erythematosus. *Pak J Med Sci.* 2017;33(1):59-64. doi:10.12669/pjms.331.11947

[12] Kanapathy A, Jaafar NR, Shaharir SS, et al. Prevalence of cognitive impairment using the Montreal Cognitive Assessment questionnaire among patients with systemic lupus erythematosus: a cross-sectional study at two tertiary centres in Malaysia. *Lupus.* 2019;28(7):854-861. doi:10.1177/0961203319852153

[13] Nishimura K, Omori M, Katsumata Y, et al. Neurocognitive impairment in corticosteroid-naïve patients with active systemic lupus erythematosus: a prospective study. *J Rheumatol.* 2015;42:441-448. doi:10.3899/jrheum.140659

[14] Masadeh A, Nofal BM, Masa'deh R. Effect of Benson relaxation response technique on the quality of life among patients with systemic lupus erythematosus: quasi-experimental study. *Lupus Sci Med.* 2025;12(1):e001301. doi:10.1136/lupus-2024-001301

[15] Tanaka Y, Miyazaki Y, Hirata S, et al. Impact of quality of life on overall work productivity impairment and activity impairment of patients with systemic lupus erythematosus: the PEONY study. *Lupus Sci Med.* 2025;12(1):e001291. doi:10.1136/lupus-2024-001291

[16] Alzheimer's Disease International. World Alzheimer Report 2021: Journey through the diagnosis of dementia. London: Alzheimer's Disease International; 2021. Available from: https://www.alzint.org/resource/world-alzheimer-report-2021/

[17] Velez-Coto M, Rute-Perez S, Perez-Garcia M, Caracuel A. Unemployment and general cognitive ability: A review and meta-analysis. *J Econ Psychol.* 2021;87:102430. doi:10.1016/j.joep.2021.102430

[18] Monahan R, Middelkoop H, Beaart-van de Voorde L, et al. High prevalence but low impact of cognitive dysfunction on quality of life in patients with lupus and neuropsychiatric symptoms. *Arthritis Care Res (Hoboken)*. 2022;75(5). doi:10.1002/acr.24904

[19] Mackay M. Lupus brain fog: A biologic perspective on cognitive impairment, depression, and fatigue in systemic lupus erythematosus. *Immunol Res*. 2015;63:26-37. doi:10.1007/s12026-015-8716-3

[20] Plantinga L, Tift B, Dunlop-Thomas C, Lim S, Bowling C, Drenkard C. Geriatric assessment of physical and cognitive functioning in a diverse cohort of systemic lupus erythematosus patients: A pilot study. *Arthritis Care Res (Hoboken)*. 2018;70(10):1469-1477. doi:10.1002/acr.23507

[21] Meara A, Davidson N, Steigelman H, et al. Screening for cognitive impairment in SLE using the Self-Administered Gerocognitive Exam. *Lupus*. 2018;27:1363-1367. doi:10.1177/0961203318759429

[22] Panopalis P, Julian L, Yazdany J, et al. Impact of memory impairment on employment status in persons with systemic lupus erythematosus. *Arthritis Rheumatol*. 2007;57(8):1453-1460. doi:10.1002/art.23090

[23] Harrison M. Thinking, memory, and lupus: What we know and what we can do. Published 2006. Accessed April 16, 2025. https://www.hss.edu/conditions_thinking-memory-lupus.asp

[24] Whittall-Garcia L, Urowitz MB, Gladman DD, et al. 1303 The new EULAR/ACR 2019 SLE classification criteria: a predictor of long-term outcomes. *Lupus Sci Med*. 2021;8. doi:10.1016/j.semarthrit.2022.152103

[25] Duarte-Garcia A, Romero-Diaz J, Juarez S, et al. Disease activity, autoantibodies, and inflammatory molecules in serum and cerebrospinal fluid of patients with systemic lupus erythematosus and cognitive dysfunction. *PLoS One*. 2018;13(5):e0196487. doi:10.1371/journal.pone.0196487

Chapter 10

[1] Harrison M. Thinking, memory, and lupus: What we know and what we can do. Hospital for Special Surgery. Published 2006. Accessed April 16, 2025. https://www.hss.edu/conditions_thinking-memory-lupus.asp

[2] Wallace DJ, Hahn BH, eds. *Dubois' Lupus Erythematosus*. 7th ed. Philadelphia, PA: Lippincott Williams & Wilkins; 2007. Accessed April 16, 2025. https://www.wolterskluwer.com/en/solutions/ovid/dubois-lupus-erythematosus-3484

[3] Kello N, Anderson E, Diamond B. Cognitive dysfunction in systemic lupus erythematosus: A case for initiating trials. *Arthritis Rheumatol*. 2019;71(9):1413-1425. doi:10.1002/art.40933

[4] Simon S, Yokomizo J, Bottino C. Cognitive intervention in amnestic mild cognitive impairment: a systematic review. *Neurosci Biobehav Rev*. 2012;36(4):1163–1178. doi:10.1016/j.neubiorev.2012.01.007

[5] Cummings J, Aisen P, Apostolova LG, Atri A, Salloway S, Weiner M. Aducanumab: Appropriate use recommendations. *J Prev Alzheimers Dis*. 2021;8(4):398-410. doi:10.14283/jpad.2021.41

[6] Boumpas D. Management of neuropsychiatric SLE. *Lupus Sci Med*. 2022;9(Suppl 1):A8.2. doi:10.1136/lupus-2022-lupus21century.19

[7] Chandler MJ, Parks AC, Marsiske M, Rotblatt LJ, Smith GE. Everyday impact of cognitive interventions in mild cognitive impairment: A systematic review and meta-analysis. *Neuropsychol Rev*. 2017;26(3):225-251. doi:10.1007/s11065-016-9330-4

[8] Lupus Foundation of America. How lupus affects memory. Lupus Foundation of America. https://www.lupus.org/resources/how-lupus-affects-memory. Accessed April 16, 2025.

[9] Mani A, Shenavandeh S, Sepehrtaj SS, Javadpour A. Memory and learning functions in patients with systemic lupus erythematosus: a

neuropsychological case-control study. *Egypt Rheumatol.* 2015;37(4)(suppl):S13-S17. doi:10.1016/j.ejr.2015.02.004

[10] Birt JA, Hadi MA, Sargalo N, et al. Patient experiences, satisfaction, and expectations with current systemic lupus erythematosus treatment: results of the SLE-UPDATE Survey. *Rheumatol Ther.* 2021. doi:10.1007/s40744-021-00328-6

[11] Khoury LE, Zarfeshani A, Diamond B. Using the mouse to model human diseases: cognitive impairment in systemic lupus erythematosus. *J Rheumatol.* 2020;47(7):1145-1149. doi:10.3899/jrheum.200410

[12] Harrison MJ, Morris KA, Horton R, et al. Results of intervention for lupus patients with self-perceived cognitive difficulties. *Neurology.* 2005;65(8):1325-1327. doi:10.1212/01.wnl.0000180938.69146.5e

[13] Hong YJ, Jang EH, Hwang J, Roh JH, Lee JH. The efficacy of cognitive intervention programs for mild cognitive impairment: a systematic review. *Curr Alzheimer Res.* 2015;12(6):undefined. doi:10.2174/1567205012666150530201636

[14] Reijnders J, van Heugten C, van Boxtel M. Cognitive interventions in healthy older adults and people with mild cognitive impairment: a systematic review. *Ageing Res Rev.* 2013;12(1):263–275. doi:10.1016/j.arr.2012.07.003

[15] Panopalis P, Julian L, Yazdany J, et al. Impact of memory impairment on employment status in persons with systemic lupus erythematosus. *Arthritis Rheum.* 2007;57(8):1453-1460. doi:10.1002/art.23090

[16] Yang HL, Chan PT, Chang PC, et al. Memory-focused interventions for people with cognitive disorders: a systematic review and meta-analysis of randomized controlled studies. *Int J Nurs Stud.* 2018;78:44-51. doi:10.1016/j.ijnurstu.2017.08.005

[17] Wang C, Yu JT, Wang HF, et al. Non-pharmacological interventions for patients with mild cognitive impairment: a meta-analysis of randomized controlled trials of cognition-based and exercise

interventions. *J Alzheimers Dis.* 2014;42(2):663-678. doi:10.3233/JAD-140660

[18] Fleck C, Corwin M. Evidence-based decisions: memory intervention for individuals with mild cognitive impairment. *EBP Briefs.* 2013;8:1-14. Available at: https://www.pearsonassessments.com/content/dam/school/global/clinical/us/assets/ebp-briefs/EBPV8A4.pdf

[19] Folkerts A, Roheger M, Frankline J, Middelstadt J, Kalbe E. Cognitive interventions in patients with dementia living in long-term care facilities: systematic review and meta-analysis. *Arch Gerontol Geriatr.* 2017;73:204-221. Available at: https://www.researchgate.net/publication/318770190_Cognitive_interventions_in_patients_with_dementia_living_in_long-term_care_facilities_Systematic_review_and_meta-analysis

[20] Cotelli M, Manenti R, Gobbi E, et al. Cognitive telerehabilitation in mild cognitive impairment, Alzheimer's disease and frontotemporal dementia: a systematic review. *Int J Telemed eHealth.* 2017. doi:10.1177/1357633X17740390

[21] Lesher AI, Landis S, Stroud C, Downey A, eds. *Preventing Cognitive Decline and Dementia: A Way Forward.* Washington, DC: The National Academies Press; 2017. doi:10.17226/24782

[22] Masley SM. *The Better Brain Solution: How to Start Now at Any Age and Prevent Insulin Resistance of the Brain, Sharpen Cognitive Function, and Avoid Memory Loss.* New York: Alfred A. Knopf; 2018. Available at: https://a.co/d/9zg2Ixo

[23] Mueller KD. A review of computer-based cognitive training for individuals with mild cognitive impairment and Alzheimer's disease. *Perspect ASHA Spec Interest Groups.* 2019;1(1):47. doi:10.1044/persp1.SIG2.47

[24] Klimova B, Maresova P. Computer-based training programs for older people with mild cognitive impairment and/or dementia. *Front Hum Neurosci.* 2017;11:262. doi:10.3389/fnhum.2017.00262

[25] Hill NT, Mowszowski L, Naismith SL, et al. Computerized cognitive training in older adults with mild cognitive impairment or dementia: a systematic review and meta-analysis. *Am J Psychiatry*. 2016. Available at: https://doi.org/10.1176/appi.ajp.2016.16030360

[26] Coyle H, Traynor V, Solowij N. Computerized and virtual reality cognitive training for individuals at high risk of cognitive decline: Systematic review of the literature. *Am J Geriatr Psychiatry*. 2015;23(4):335-359. doi:10.1016

[27] Gary R, Paul S, Corwin EL, et al. Exercise and cognitive training as a strategy to improve neurocognitive outcomes in heart failure: a pilot study. *Am J Geriatr Psychiatry*. 2019;27(8):809-819. doi:10.1016/j.jagp.2019.01.211

[28] Edwards JD, Xu H, Clark DO, Guey LT, Ros LA, Unverzagt FW. Speed of processing training results in lower risk of dementia. *Alzheimers Dement (N Y)*. 2017;3(4):603–611. doi:10.1016/j.trci.2017.09.002

[29] Martin CB, Hong B, Newsome RN, et al. A smartphone intervention that enhances real-world memory and promotes differentiation of hippocampal activity in older adults. *Proc Natl Acad Sci U S A*. 2022;119(51):e2214285119. doi:10.1073/pnas.2214285119

[30] Hill NT, Mowszowski L, Naismith SL, et al. Computerized cognitive training in older adults with mild cognitive impairment or dementia: a systematic review and meta-analysis. *Am J Psychiatry*. 2016. doi:10.1176/appi.ajp.2016.16030360

Chapter 11

[1] Langa KM, Levine DA. The diagnosis and management of mild cognitive impairment: a clinical review. *JAMA*. 2014;312(23):2551-2561. doi:10.1001/jama.2014.13806

[2] Peterson RC, Lopez O, Armstrong MJ, et al. Practice guideline update: Mild cognitive impairment. Report of the Guideline Development, Dissemination, and Implementation Subcommittee of the American

Academy of Neurology. *Neurology.* 2017;doi:10.1212/WNL.0000000000004826

[3] Fujita Y, Fukui S, Ishida M, et al. Reversible cognitive dysfunction in elderly-onset systemic lupus erythematosus, successfully treated with aggressive immunosuppressive therapy. *Intern Med.* 2018;57(20):3025-3028. doi:10.2169/internalmedicine.0934-18

[4] Denburg SD, Carbotte RM, Denburg JA. Corticosteroids and neuropsychological functioning in patients with systemic lupus erythematosus. *Arthritis Rheum.* 1994;37:1311-1320. doi:10.1002/art.1780370907

[5] Caceres V. Pinpoint cognitive dysfunction in patients with lupus. *The Rheumatologist.* Published 2012. Accessed April 16, 2025. https://www.the-rheumatologist.org/article/pinpoint-cognitive-dysfunction-in-patients-with-lupus/

[6] Popescu A, Kao AH. Neuropsychiatric systemic lupus erythematosus. *Curr Neuropharmacol.* 2011;9(3):449-457. doi:10.2174/157015911796557984

[7] Tao R, Dhanasekaran P, Tay S, Mak A. Mathematical processing is affected by daily but not cumulative glucocorticoid dose in patients with systemic lupus erythematosus. *Rheumatology.* 2020;59(12):2534-2543. doi:10.1093/rheumatology/keaa002

[8] Kozora E, West SG, Kotzin BL, et al. Magnetic resonance imaging abnormalities and cognitive deficits in systemic lupus erythematosus patients without overt central nervous system disease. *Arthritis Rheumatol.* 1998;41:41-47. doi:10.1002/1529-0131(199801)41:1<41::AID-ART6>3.0.CO;2-7

[9] Appenzeller S, Rondina JM, Li LM, et al. Cerebral and corpus callosum atrophy in systemic lupus erythematosus. *Arthritis Rheumatol.* 2005;52(9):2783-2789. doi:10.1002/art.21271

[10] Butt B, Farman S, Khan S, Saeed MA, Ahmad NM. Cognitive dysfunction in patients with Systemic Lupus Erythematosus. *Pak J Med Sci.* 2017;33(1):59-64. doi:10.12669/pjms.331.11947

[11] Harrison M. Thinking, memory, and lupus: What we know and what we can do. Hospital for Special Surgery. Published 2006. Accessed April 16, 2025. https://www.hss.edu/conditions_thinking-memory-lupus.asp

[12] Liu S, Cheng Y, Zhao Y, et al. Clinical factors associated with brain volume reduction in systemic lupus erythematosus patients without major neuropsychiatric manifestations. *Front Psychiatry.* 2018;9:8. doi:10.3389/fpsyt.2018.00008

[13] Sood A, Al S, Raji M. Cognitive impairment in elderly patients with rheumatic disease and the effect of disease-modifying anti-rheumatic drugs. *Clin Rheumatol.* 2021;40(2):491-498. doi:10.1007/s10067-020-05372-1

[14] Magro-Checa C, Zirkzee EJ, Huizinga TW, Steup-Beekman GM. Management of neuropsychiatric systemic lupus erythematosus: current approaches and future perspectives. *Drugs.* 2016;76(4):459-483. doi:10.1007/s40265-015-0534-3

[15] Faria R, Goncalves J, Dias R. Neuropsychiatric systemic lupus erythematosus involvement: toward a tailored approach to our patients? *Rambam Maimonides Med J.* 2017;8(1):e0001. doi:10.5041/RMMJ.10276

[16] Piga M, Peltz MT, Montaldo C, et al. Twenty-year brain magnetic resonance imaging follow-up study in systemic lupus erythematosus: factors associated with accrual of damage and central nervous system involvement. *Autoimmun Rev.* 2015;14(6):510-516. doi:10.1016/j.autrev.2015.01.010

[17] McLaurin EY, Holliday SL, Williams P, Brey RL. Predictors of cognitive dysfunction in patients with systemic lupus erythematosus. *Neurology.* 2005;64(2):297-303. doi:10.1212/01.WNL.0000149640.78684.EA

[18] Pavlakis P. Rheumatologic disorders and the nervous system. *Neurology of Systemic Disease.* 2020;26(3):591–610. doi:10.1212/CON.0000000000000856

[19] Song K, Han HJ, Kim S, Kwon J. Thymosin beta 4 attenuates PrP(106-126)-induced human brain endothelial cells dysfunction. *Eur J Pharmacol.* 2020;869:172891. doi:10.1016/j.ejphar.2019.172891

[20] Kang HK, Liu M, Datta SK. Low-dose peptide tolerance therapy of lupus generates plasmacytoid dendritic cells that cause expansion of autoantigen-specific regulatory T cells and contraction of inflammatory Th17 cells. *J Immunol.* 2007;178(12):7849-7858. doi:10.4049/jimmunol.178.12.7849

[21] Barraclough M, McKie S, Parker B, et al. Altered cognitive function in systemic lupus erythematosus and associations with inflammation and functional and structural brain changes. *Ann Rheum Dis.* 2019;78:934-940. doi:10.1136/annrheumdis-2018-214677

[22] Fesharaki-Zadeh A, Lowe N, Arnsten A. Clinical experience with the α2A-adrenoceptor agonist, guanfacine, and N-acetylcysteine for the treatment of cognitive deficits in "Long-COVID19." *Neuroimmunol Rep.* 2023;3:100154. doi:10.1016/j.nerep.2022.100154

[23] Lopez P, de Paz B, Rodriguez-Carrio J, et al. Th17 responses and natural IgM antibodies are related to gut microbiota composition in systemic lupus erythematosus patients. *Sci Rep.* 2016;6:24072. doi:10.1038/srep24072

[24] Mani A, Shenavandeh S, Sepehrtaj SS, Javadpour. Memory and learning functions in patients with systemic lupus erythematosus: a neuropsychological case-control study. *Egypt Rheumatol.* 2015;37(4):S13-S17. doi:10.1016/j.ejr.2015.02.004

[25] Tjensvoll A, Lauvsnes M, Zetterberg H, et al. Neurofilament light is a biomarker of brain involvement in lupus and primary Sjögren's syndrome. *J Neurol.* 2021;268(4):1385-1394. doi:10.1007/s00415-020-10290-y

[26] Nicolaou O, Kousios A, Hadisavvas A, et al. Biomarkers of systemic lupus erythematosus identified using mass spectrometry-based proteomics: a systematic review. *J Cell Mol Med.* 2017;21(5):993-1012. doi:10.1111/jcmm.13031

[27] Murray S, Yazdany J, Kaiser R, Criswell L. Cardiovascular disease and cognitive dysfunction in systemic lupus erythematosus. *Arthritis Care Res (Hoboken).* 2012;64(9):1328-1333. doi:10.1002/acr.21691

[28] Gerosa M, Poletti B, Pregnolato F, et al. Antiglutamate receptor antibodies and cognitive impairment in primary antiphospholipid syndrome and systemic lupus erythematosus. *Front Immunol.* 2016;7:5. doi:10.3389/fimmu.2016.00005

[29] Balajkova V, Olejarova M, Moravcova R, et al. Is serum TWEAK a useful biomarker of neuropsychiatric systemic lupus erythematosus. *Physiol Res.* 2020;69(2):339-346. doi:10.33549/physiolres.934308

[30] Wang JB, Li H, Wang LL, et al. Role of IL-1β, IL-6, IL-8, and IFN in pathogenesis of central nervous system neuropsychiatric systemic lupus erythematosus. *Int J Clin Exp Med.* 2015;8(9):16658-16663. PMID: 26629199

[31] Schretlen DJ, Inscore AB, Jinnah HA, Rao V, Gordon B, Pearlson GD. Serum uric acid and cognitive function in community-dwelling older adults. Neuropsychology. 2007 Jan;21(1):136-40. doi: 10.1037/0894-4105.21.1.136. PMID: 17201536.

[32] Diaz-Gallo LM, Oke V, Lundstrom Em, et al. Four systemic lupus erythematosus subgroups, defined by autoantibodies status, differ regarding HLA-DRB1 genotype associations and immunological and clinical manifestations. *Arthritis Care Res (Hoboken).* 2021. doi:10.1002/acr2.11343

[33] Vollenhoven R. Conceptual framework for defining disease modification in systemic lupus erythematosus: a call for formal criteria. *Lupus Sci Med.* 2022;9(1):e000634. doi:10.1136/lupus-2021-000634

[34] Northcott M, Jones S, Koelmeyer R, et al. Type 1 interferon status in systemic lupus erythematosus: a longitudinal analysis. *Lupus Sci Med.* 2021;9(1):e000625. doi:10.1136/lupus-2021-000625

[35] Nehar-Belaid D, Hong S, Marches R, et al. Mapping systemic lupus erythematosus heterogeneity at the single-cell level. *Nat Immunol.* 2020;21(9):1094-1106. Available at: https://www.ncbi.nlm.nih.gov/projects/gap/cgi-bin/study.cgi?study_id=phs002048.v2.p1

[36] Cristina Arriens, Jonathan D. Wren, Melissa E. Munroe, Chandra Mohan, Systemic lupus erythematosus biomarkers: the challenging quest, *Rheumatology*, Volume 56, Issue suppl_1, April 2017, Pages i32–i45, https://doi.org/10.1093/rheumatology/kew407

[37] Cao ZY, Wang N, Jia JT, et al. Abnormal topological organization in systemic lupus erythematosus: a resting-state functional magnetic resonance imaging analysis. *Brain Imaging Behav.* 2020. doi:10.1007/s11682-019-00228-y

[38] Atkinson J, Yu CY. Complement in SLE. *BMJ.* 2021;8(1). doi:10.1186/ar586

[39] Banchereau J, Pascual V. Type I interferon in systemic lupus erythematosus and other autoimmune diseases. *Immunity.* 2006;25(3):383-392. doi:10.1016/j.immuni.2006.08.010

[40] Cagnoli P, Haris RE, Frechtling D, et al. Reduced insular glutamine and N-acetylaspartate in systemic lupus erythematosus. *Acad Radiol.* 2013;20(10):1286-1296. doi:10.17615/nvxn-a725

[41] Zhang Z, Wang Y, Shen Z, et al. The neurochemical and microstructural changes in the brain of systemic lupus erythematosus patients: a multimodal MRI study. *Sci Rep.* 2016;6:19026. doi:10.1038/srep19026

[42] Duarte-Garcia A, Romero-Diaz J, Juarez S, et al. Disease activity, autoantibodies, and inflammatory molecules in serum and cerebrospinal fluid of patients with systemic lupus erythematosus and cognitive

dysfunction. *PLoS One*. 2018;13(5):e0196487. doi:10.1371/journal.pone.0196487

[43] Fernandez H, Cevallos A, Sotomayor RJ, Naranjo-Saltos F, Orces DM, Basantes E. Mental disorders in systemic lupus erythematosus: a cohort study. *Rheumatol Int*. 2019;39:2117-2125. doi:10.1007/s00296-019-04423-4

[44] Kello N, Anderson E, Diamond B. Cognitive dysfunction in systemic lupus erythematosus: a case for initiating trials. *Arthritis Rheumatol*. 2019;71(9):1413-1425. doi:10.1002/art.40933

[45] Yue R, Gurung I, Long XX, Xian JY, Peng XB. Prevalence, involved domains, and predictor of cognitive dysfunction in systemic lupus erythematosus. *Lupus*. 2020;29(13):1743-1751. doi:10.1177/0961203320958061

[46] Hirohata S, Arinuma Y, Yanagida T, Yoshio T. Blood-brain barrier damages and intrathecal synthesis of anti-N-methyl-D-aspartate receptor NR2 antibodies in diffuse psychiatric/neuropsychological syndromes in systemic lupus erythematosus. *Arthritis Res Ther*. 2014;16(2):R77. doi:10.1186/ar4518

[47] Northcott M, Jones S, Koelmeyer R, et al. Type 1 interferon status in systemic lupus erythematosus: a longitudinal analysis. *Lupus Sci Med*. 2021;9(1):e000625. doi:10.1136/lupus-2021-000625

[48] Doria A, Gatto M, Zen M, Iaccarino L, Punzi L. Optimizing outcome in SLE: treating-to-target and definition of treatment goals. *Autoimmun Rev*. 2014;13(7):770-777. doi:10.1016/j.autrev.2014.01.055

[49] Thurman J, Serkova N. Non-invasive imaging to monitor lupus nephritis and neuropsychiatric systemic lupus erythematosus. *F1000Res*. 2015;4:153. doi:10.12688/f1000research.6587.2

[50] Moore E, Huang MW, Putterman C. Advances in the diagnosis, pathogenesis, and treatment of neuropsychiatric systemic lupus erythematosus. *Rheumatology*. 2020;32(2):302-312. doi:10.1097/BOR.0000000000000682

[51] Ho RC, Thiaghu C, Ong H, et al. A meta-analysis of serum and cerebrospinal fluid autoantibodies in neuropsychiatric systemic lupus erythematosus. *Autoimmun Rev.* 2016;15(2):124-138. doi:10.1016/j.autrev.2015.10.003

[52] Wiseman SJ, Bastin ME, Hamilton IF, et al. Fatigue and cognitive function in systemic lupus erythematosus: associations with white matter microstructural damage. A diffusion tensor MRI study and meta-analysis. *Lupus.* 2017;26(6):588-597. doi:10.1177/0961203316668417

[53] Lapa A, Postal M, Sinicato N, et al. S100B is associated with cognitive impairment in childhood-onset systemic lupus erythematosus patients. *Lupus.* 2017;26(5):478-483. doi:10.1177/0961203317691374

[54] Grajales CM, Barraclough ML, Diaz-Martinez JP, et al. v402 Serum cytokine profiling reveals elevated levels of S100A8/A9 and MMP-9 in systemic lupus erythematosus patients with cognitive impairment independently of disease activity and inflammatory markers. *Lupus Sci Med.* 2022;9(Suppl 1):A14. doi:10.1136/lupus-2022-lupus21century.14

[55] Riancho-Zarrabeitia L, Martinez-Toboada VM, Rua-Figueroa I, et al. Do all antiphospholipid antibodies confer the same risk for major organ involvement in systemic lupus erythematosus patients? *Clin Exp Rheumatol.* 2020 Aug 7. doi:10.55563/clinexprheumatol/9kxexc

[56] Anderson E, Jin Y, Goodwin S, et al. The association of interferon-alpha with kynurenine/tryptophan pathway activation in systemic lupus erythematosus. *American College of Rheumatology Convergence Meeting Abstracts.* 2020. https://acrabstracts.org/abstract/the-association-of-interferon-%CE%B1-with-kynurenine-tryptophan-pathway-activation-in-systemic-lupus-erythematosus/

[57] Jenks SA, Cashman KS, Zumaquero E, Boss JM, Lund FE, Sanz I. Distinct effector B cells induced by unregulated Toll-like receptor 7 contribute to pathogenic responses in systemic lupus erythematosus. *Immunity.* 2018;49(4):725-739. doi:10.1016/j.immuni.2018.08.015

[58] Rose T, Dorner T. Drivers of the immunopathogenesis in systemic lupus erythematosus. *Best Pract Res Clin Rheumatol.* 2017;31(3):321-333. doi:10.1016/j.berh.2017.09.007

[59] Gergianaki I, Bortoluzzi A, Bertsias G. Update on the epidemiology, risk factors, and disease outcomes of systemic lupus erythematosus. *Best Pract Res Clin Rheumatol.* 2018;32(2):188-205. doi:10.1016/j.berh.2018.09.004

[60] Felten R, Dervovic E, Chasset F, et al. The 2018 pipeline of targeted therapies under clinical development for systemic lupus erythematosus: a systematic review of trials. *Autoimmun Rev.* 2018;17(8):781–790. doi:10.1016/j.autrev.2018.02.011

[61] Mike EV, Makinde HM, Der E, et al. Neuropsychiatric systemic lupus erythematosus is dependent on sphingosine-1-phosphate signaling. *Front Immunol.* 2018;9:2198. doi:10.3389/fimmu.2018.02189

[62] Khoury LE, Zarfeshani A, Diamond B. Using the mouse to model human diseases: cognitive impairment in systemic lupus erythematosus. *J Rheumatol.* 2020;47(7):1145-1149. doi:10.3899/jrheum.200410

[63] Nestor J, Arinuma Y, Huerta TS, et al. Lupus antibodies induce behavioral changes mediated by microglia and blocked by ACE inhibitors. *J Exp Med.* 2018;215(10):2554-2566. doi:10.1084/jem.20180776

[64] Morara S. Brain inflammation: the case of CGRP. *Neuropeptides.* 2017;65:133-134. doi:10.1016/j.npep.2017.02.019

[65] Nikolopoulos D, Fanouriakis A, Boumpas D. Update on the pathogenesis of central nervous system lupus. *Curr Opin Rheumatol.* 2019;31(6):669-677. doi:10.1097/BOR.0000000000000655

[66] Takahashi J, Ueta Y, Yamada D, et al. Intracerebroventricular administration of oxytocin and intranasal administration of the oxytocin derivative improve β-amyloid peptide (25–35)-induced memory impairment in mice. *Neuropsychopharmacol Rep.* 2022. doi:10.1002/npr2.12292

[67] Leiter O, Zhuo Z, Rust R, et al. Selenium mediates exercise-induced adult neurogenesis and reverses learning deficits induced by hippocampal injury and aging. *Cell Metab.* 2022. doi:10.1016/j.cmet.2022.01.005

[68] Arriens C, Hynana LS, Lerman RH, Karp DR, Mohan C. Placebo-controlled randomized clinical trial of fish oil's impact on fatigue, quality of life, and disease activity in systemic lupus erythematosus. *J Nutr.* 2015;14(1):82. doi:10.1186/s12937-015-0068-2

[69] Masley SM. *The Better Brain Solution: How to Start Now at Any Age and Prevent Insulin Resistance of the Brain, Sharpen Cognitive Function, and Avoid Memory Loss.* New York: Alfred A. Knopf; 2018. https://a.co/d/9zg2Ixo

[70] Morris G, Berk M, Walder K, Maes M. Central pathways causing fatigue in neuro-inflammatory autoimmune illnesses. *BMC Med.* 2015;13:28. doi:10.1186/s12916-014-0259-2

[71] Huang HA. Multicentre, randomised, double blind, parallel design, placebo controlled study to evaluate the efficacy and safety of Uthever (NMN supplement), an orally administered supplementation in middle aged and older adults. *Front Aging.* 2022;3:851698. doi:10.3389/fragi.2022.851698

[72] Orefice B, Ceccarelli F, Barbati C, et al. Caffeine intake influences disease activity and clinical phenotype in systemic lupus erythematosus patients. *Lupus.* 2020. https://journals.sagepub.com.

[73] Chen BD, Jia XM, Xu JY, Zhao LD, Ji JY, Wu BX, Ma Y, Li H, Zuo XX, Pan WY, Wang XH, Ye S, Tsokos GC, Wang J, Zhang X. An autoimmunogenic and proinflammatory profile defined by the gut microbiota of patients with untreated systemic lupus erythematosus. *Arthritis Rheumatol.* 2021;73(2):232-243. doi:10.1002/art.41511

[74] Xiang S, Qu Y, Qian S, et al. Association between systemic lupus erythematosus and disruption of gut microbiota: a meta-analysis. *Lupus Sci Med.* 2022;9(1). doi:10.1136/lupus-2021-000599

[75] Tedeschi S, Barbhaiya M, Sparks J, et al. Dietary patterns and risk of systemic lupus erythematosus in women. *Lupus.* 2020;29(1):67–73. doi:10.1177/0961203319888791

[76] Baker LD, Manson JE, Rapp SR, et al. Effects of cocoa extract and a multivitamin on cognitive function: A randomized clinical trial. *Alzheimers Dement.* 2022. doi:10.1002/alz.12767

[77] Tay SH, Ho CS, Ho RC, Mak A. 25-Hydroxyvitamin D3 deficiency independently predicts cognitive impairment in patients with systemic lupus erythematosus. *PLoS One.* 2015;10(12). doi:10.1371/journal.pone.0144149

[78] Hussein H, Daker L, Fouad N, Elamir A, Mohamed S. Does vitamin D deficiency contribute to cognitive dysfunction in patients with systemic lupus erythematosus? *Innov Clin Neurosci.* 2018;15(9-10):25-29. doi: 30588363

[79] Yan LJ, Wu P, Gao DM, et al. The impact of vitamin D on cognitive dysfunction in mice with systemic lupus erythematosus. *Med Sci Monit.* 2019;25:4716-4722. doi:10.12659/MSM.915355

[80] Dean T. Study reveals new insights into the link between sunlight exposure and kidney damage in lupus. *Dartmouth Geisel School of Medicine.* Published 2021. Accessed April 18, 2025. https://geiselmed.dartmouth.edu/news/2021/study-reveals-new-insights-into-the-link-between-sunlight-exposure-and-kidney-damage-in-lupus/

[81] Greiling TM, Dehner C, Chen X, et al. Commensal orthologs of the human autoantigen Ro60 as triggers of autoimmunity in lupus. *Sci Transl Med.* 2018;10(434):eaan2306. doi:10.1126/scitranslmed.aan2306

[82] Beaudry-Richard A, Abdelhak A, Saloner R, Sacco S, Montes SC, Oertel FC, et al. Vitamin B12 Levels Association with Functional and Structural Biomarkers of Central Nervous System Injury in Older Adults. *Ann Neurol.* 2025. doi:10.1002/ana.27200

[83] Ramos A, Garrison V, Koop D, Rooney W, Foundas A, Bourdette D. Polyphenon E, a green tea extract, increases brain NAA levels in MS: a

pilot six-month open label study. *Neurology*. 2012;78(1):P03.050. doi:10.1212/WNL.78.1_supplement.P03.050

[84] Shibata S, Noguchi-Shinohara M, Shima A, et al. Green tea consumption and cerebral white matter lesions in community-dwelling older adults without dementia. *npj Sci Food*. 2025;9(2). doi:10.1038/s41538-024-00364-w

[85] Pan M-H, Lai C-S, Wang H, Lo C-Y, Ho C-T, Li S. Black tea in chemo-prevention of cancer and other human diseases. *Food Sci Hum Wellness*. 2013;2(1):12-21. doi:10.1016/j.fshw.2013.03.004

[86] Li G, Wu Q, Wang C, Deng P, Li J, Zhai Z, Li Y. Curcumin reverses cognitive deficits through promoting neurogenesis and synapse plasticity via the upregulation of PSD95 and BDNF in mice. *Sci Rep*. 2025;15(1):1135. doi:10.1038/s41598-024-82571-9

[87] Chang C, Zhao Y, Song G, She K. Resveratrol protects hippocampal neurons against cerebral ischemia-reperfusion injury via modulating JAK/ERK/STAT signaling pathway in rats. *J Neuroimmunol*. 2018;315:9-14. doi:10.1016/j.jneuroim.2017.11.015

[88] Alesci A, Nicosia N, Fumia A, Giorgianni F, Santini A, Cicero N. Resveratrol and immune cells: a link to improve human health. *Molecules*. 2022;27(2):424. doi:10.3390/molecules27020424

[89] McGrattan AM, McEvoy CT, McGuinness B, McKinley MC, Woodside JV. Effect of dietary interventions in mild cognitive impairment: a systematic review. *Br J Nutr*. 2018;120(12):1388-1405. doi:10.1017/S0007114518002945

[90] Butler, M.J., Deems, N.P., Muscat, S., Butt, C.M., Belury, A., & Barrientos, R.M. (2021). Dietary DHA prevents cognitive impairment and inflammatory gene expression in aged male rats fed a diet enriched with refined carbohydrates. *Brain, Behavior, and Immunity,* 98, 198-209. https://doi.org/10.1016/j.bbi.2021.08.214.

[91] Gilley KN, Fenton JI, Zick SM, Li K, Wang L, Marder W, McCune WJ, Jain R, Herndon-Fenton S, Hassett AL, Barbour KE, Pestka JJ, Somers EC. Serum fatty acid profiles in systemic lupus erythematosus

and patient reported outcomes: The Michigan Lupus Epidemiology & Surveillance (MILES) Program. *Front Immunol*. 2024;15:1459297. doi:10.3389/fimmu.2024.1459297

[92] Ali RA, Gandhi AA, Dai L, Weiner J, Estes SK, Yalavarthi S, Gockman K, Sun D, Knight JS. Antineutrophil properties of natural gingerols in models of lupus. JCI Insight. 2021 Feb 8;6(3):e138385. doi: 10.1172/jci.insight.138385. PMID: 33373329; PMCID: PMC7934838.

[93] Kim YS, Won YJ, Lim BG, Min TJ, Kim YH, Lee IO. Neuroprotective effects of magnesium l-threonate in a hypoxic zebrafish model. *BMC Neurosci*. 2020;21(1):29. doi:10.1186/s12868-020-00580-6

[94] Dave N, Judd JM, Decker A, Winslow W, Sarette P, Villarreal Espinosa O, Tallino S, Bartholomew SK, Bilal A, Sandler J, McDonough I, Winstone JK, Blackwood EA, Glembotski C, Karr T, Velazquez R. Dietary choline intake is necessary to prevent systems-wide organ pathology and reduce Alzheimer's disease hallmarks. *Aging Cell*. 2023;22(2):e13775. doi:10.1111/acel.13775

[95] Sciascia S, Ferrara G, Roccatello L, Rubini E, Foddai SG, Radin M, Cecchi I, et al. The interconnection between systemic lupus erythematosus and diet: unmet needs, available evidence, and guidance—a patient-driven, multistep-approach study. *Nutrients*. 2024;16(23):4132. doi:10.3390/nu16234132

[96] Li W, Qin L, Feng R, Hu G, Sun H, He Y, Zhang R. Emerging senolytic agents derived from natural products. *Mech Ageing Dev*. 2019;181:1-6. doi:10.1016/j.mad.2019.05.001

Chapter 12

[1] Klimova, B., & Maresova, P. (2017). Computer-based training programs for older people with mild cognitive impairment and/or dementia. *Frontiers in Human Neuroscience*. 11, 262. doi: *10.3389/fnhum.2017.00262*

[2] Mellow ML, Crozier AJ, Dumuid D, Wade AT, Goldsworthy MR, Dorrian J, Smith AE. (2022). How are combinations of physical activity,

sedentary behaviour and sleep related to cognitive function in older adults? A systematic review. *Exp Gerontol.* 2022 Mar;159:111698. DOI: 10.1016/j.exger.2022.111698

[3] Smith, P.J., Blumenthal, J.A., Hoffman, B.M., Cooper, H., Strauman, T.Z., Welsh-Bohmer, K., Browndyke, J.N., & Sherwood, A. (2010). Aerobic exercise and neurocognitive performance: a meta-analytic review of randomized controlled trials. *Psychosomatic Medicine,* 72(3), 239-252. doi: 10.1097/PSY.0b013e3181d14633

[4] Inoue, K., Okamoto, M., Shibato, J., Lee, M.C., Matsui, T., Rakwal, R., & soya, H. (2015). Long-term mild, rather than intense, exercise enhances adult hippocampal neurogenesis and greatly changes the transcriptomic profile of the hippocampus. *PLoS One (Public Library of Science)*, 10(6).
https://www.researchgate.net/publication/278043843_Long-Term_Mild_rather_than_Intense_Exercise_Enhances_Adult_Hippocampal_Neurogenesis_and_Greatly_Changes_the_Transcriptomic_Profile_of_the_Hippocampus

[5] Casaletto, K., Ramos-Miguel, A., VandBunte, A., Memel, M., Buchman, A., Bennett, D., & Honer, W. (2022). Late-life physical activity relates to brain tissue synaptic integrity markers in older adults. *Alzheimer's & Dementia,* https://doi.org/10.1002/alz.12530

[6] Masley, S. M. (2018). The Better Brain Solution: How to Start Now at Any Age and Prevent Insulin Resistance of the Brain, Sharpen Cognitive Function, and Avoid Memory Loss. New York: Alfred A. Knopf. https://a.co/d/9zg2Ixo

[7] Ben-Zeev, Tavor & Shoenfeld, Yehuda & Hoffman, Jay. (2022). The Effect of Exercise on Neurogenesis in the Brain. The Israel Medical Association journal : IMAJ. 24. 533-538. PMID: 35971998. https://pubmed.ncbi.nlm.nih.gov/35971998/

[8] Gorzelitz J, Trabert B, Katki H.A., Moore, S.C., Watts, E. L., & Matthews, C.E. (2022). Independent and joint associations of weightlifting and aerobic activity with all-cause, cardiovascular disease and cancer mortality in the Prostate, Lung, Colorectal and Ovarian

Cancer Screening Trial. *British Journal of Sports Medicine* Published Online First: 27 September 2022. DOI: 10.1136/bjsports-2021-105315

[9] Blackmore, D.G., Steyn, F.J., Carlisle, A., Keefe, I., Vien, K., Zhou, X., Leiter, O., Jhaveri, D., Vukovic, J., Waters, M.J., & Bartlett, P.F. (2021). An exercise "sweet spot" reverses cognitive deficits of aging by growth-hormone-induced neurogenesis. *IScience,* 24(11):103275. DOI: 10.1016/j.isci.2021.103275

[10] Ahn G, Ramsey-Goldman R. Fatigue in systemic lupus erythematosus. *Int J Clin Rheumatol.* 2012;7(2):217-227. doi:10.2217/IJR.12.4

[11] Morris G, Berk M, Walder K, Maes M. Central pathways causing fatigue in neuro-inflammatory autoimmune illnesses. *BMC Med.* 2015;13:28. doi:10.1186/s12916-014-0259-2

[12] Katz P, Julian L, Tonner MC, et al. Physical activity, obesity, and cognitive impairment among women with systemic lupus erythematosus. *Arthritis Care Res (Hoboken).* 2012;64(4):502-510. doi:10.1002/acr.21587

[13] Thomas DE. *The Lupus Encyclopedia: A Comprehensive Guide for Patients and Families.* Baltimore, MD: Johns Hopkins University Press; 2014. https://a.co/d/8Vp2WES

[14] Aghjayan S, Bournias T, Kang C, et al. Aerobic exercise improves episodic memory in late adulthood: a systematic review and meta-analysis. *Commun Med (Lond).* 2022;2:79. doi:10.1038/s43856-022-00079-7

[15] Pauluch AE, Bajpai S, Bassett DR, et al. Daily steps and all-cause mortality: a meta-analysis of 15 international cohorts. *Lancet Glob Health.* 2022;7(3):e219-e228. doi:10.1016/S2468-2667(21)00302-9

[16] Cruz B, Ahmadi M, Naismith SL, Stamatakis E. Association of daily step count and intensity with incident dementia in 78,430 adults living in the UK. *JAMA Neurol.* 2022;79(10):1059-1063. doi:10.1001/jamaneurol.2022.2672

[17] Mura G, Carta MG, Sancassiani F, Machado S, Prosperini L. Active exergames to improve cognitive functioning in neurological disabilities: a systematic review and meta-analysis. *Eur J Phys Rehabil Med.* 2018;54(3):450-462. doi:10.23736/S1973-9087.17.04680-9

[18] Anderson-Hanley C, Arciero PJ, Brickman AM, et al. Exergaming and Older Adult Cognition. *Am J Prev Med.* 2012;42(2):109-119. doi:10.1016/j.amepre.2011.10.016

[19] Yeh T, Chang K, Wu C. The active ingredient of cognitive restoration: A multicenter randomized controlled trial of sequential combination of aerobic exercise and computer-based cognitive training in stroke survivors with cognitive decline. *Arch Phys Med Rehabil.* 2019.

[20] Wang C, Yu JT, Wang HF, et al. Non-pharmacological interventions for patients with mild cognitive impairment: A meta-analysis of randomized controlled trials of cognition-based and exercise interventions. *J Alzheimers Dis.* 2014;42(2):663-678. doi:10.3233/JAD-140660

[21] Chandler MJ, Parks AC, Marsiske M, et al. Everyday impact of cognitive interventions in mild cognitive impairment: A systematic review and meta-analysis. *Neuropsychol Rev.* 2017;26(3):225-251. doi:10.1007/s11065-016-9330-4

[22] Peterson RC, Lopez O, Armstrong MJ, et al. Practice guideline update: Mild cognitive impairment. Report of the Guideline Development, Dissemination, and Implementation Subcommittee of the American Academy of Neurology. *Neurology.* 2017;doi:10.1212/WNL.0000000000004826

[23] O'Dwyer T, Durcan L, Wilson F. Exercise and physical activity in systemic lupus erythematosus: a systematic review with meta-analysis. *Semin Arthritis Rheum.* 2017;47:204-215. doi:10.1016/j.semarthrit.2017.04.003

[24] Salehinejad MA, Ghanavati E, Reinders J, et al. Sleep-dependent upscaled excitability, saturated neuroplasticity, and modulated cognition in the human brain. *Elife.* 2022;11:e69308. doi:10.7554/eLife.69308

[25] Bannai M, Kawai N. New therapeutic strategy for amino acid medicine: glycine improves the quality of sleep. *J Pharmacol Sci.* 2012;118(2):145-148. doi:10.1254/jphs.11r04fm

[26] Sumsuzzman DM, Choi J, Jin Y, Hong Y. Neurocognitive effects of melatonin treatment in healthy adults and individuals with Alzheimer's disease and insomnia: a systematic review and meta-analysis of randomized controlled trials. *Neurosci Biobehav Rev.* 2021;127:459-473. doi:10.1016/j.neubiorev.2021.04.034

[27] Leng Y, Stone KL, Yaffe K. Race differences in the association between sleep medication use and risk of dementia. *J Alzheimers Dis.* 2022. doi:10.3233/JAD-221006

[28] Holthe T, Halvorsrud L, Karterud D, et al. Usability and acceptability of technology for community-dwelling older adults with mild cognitive impairment and dementia: a systematic literature review. *Clin Interv Aging.* 2018;13:863-886. doi:10.2147/CIA.S154717

[29] Griffin RM. Lupus fog and memory problems. WebMD. https://www.webmd.com/lupus/features/lupus-fog-memory-problems#1. Accessed April 26, 2025.

[30] Harrison M. Thinking, memory, and lupus: What we know and what we can do. *Hospital for Special Surgery.* Published 2006. Accessed April 18, 2025. https://www.hss.edu/conditions_thinking-memory-lupus.asp

[31] Jordan M. *Coping with mild cognitive impairment (MCI): A guide for managing memory loss, effective brain training, and reducing the risk of dementia.* Jessica Kingsley Publishers; 2020. https://a.co/d/g6O6lLh

[32] Velikonja D, Ponsford J, Janzen S, et al. INCOG 2.0: Guidelines for cognitive rehabilitation following traumatic brain injury, Part V: Memory. *J Head Trauma Rehabil.* 2023;38(1):83-102. doi:10.1097/HTR.0000000000000837

[33] Iaccarino HF, Singer AC, Martorell AJ, et al. Gamma frequency entrainment attenuates amyloid load and modifies microglia. *Nature.* 2016;540:230-235. doi:10.1038/nature20587

[34] He Q, Colon-Motas K, Pybus A, et al. A feasibility trial of gamma sensory flicker for patients with prodromal Alzheimer's disease. *Alzheimers Dement (N Y)*. 2021;7(1):e12178. doi:10.1002/trc2.12178

[35] Pérez-González M, Badesso S, Lorenzo E, et al. Identifying the main functional pathways associated with cognitive resilience to Alzheimer's disease. Int J Mol Sci. 2021;22(17):9120. doi:10.3390/ijms22179120

[36] Weaver DF. β-Amyloid is an Immunopeptide and Alzheimer's is an Autoimmune Disease. *Curr Alzheimer Res*. 2021;18(11):849-857. doi:10.2174/1567205018666211202141650

[37] Takahashi J, Ueta Y, Yamada D, Sasaki-Hamada S, Iwai T, Akita T, et al. Intracerebroventricular administration of oxytocin and intranasal administration of the oxytocin derivative improve β-amyloid peptide (25–35)-induced memory impairment in mice. *Neuropsychopharmacol Rep*. 2022;42(4):492-501. doi:10.1002/npr2.12292

[38] McGrath H. Ultraviolet-A1 irradiation therapy for systemic lupus erythematosus. *Lupus*. 2017;26(11):1241-1251. doi:10.1177/0961203317707064

[39] Yuen H, Cunningham M. Optimal management of fatigue in patients with systemic lupus erythematosus: a systematic review. *Ther Clin Risk Manag*. 2014;10:775-786. doi:10.2147/TCRM.S56063

[40] Michael E, Covarrubias LS, Leong V, Kourtzi Z. Learning at your brain's rhythm: Individualized entrainment boosts learning for perceptual decisions. *Cereb Cortex*. 2022;1-13. doi:10.1093/cercor/bhac426

[41] Sideroff S, Wellisch D, Yarema V. A neurotherapy protocol to remediate cognitive deficits after adjuvant chemotherapy: a pilot study. *J Complement Integr Med*. 2022;20(2). doi:10.1515/jcim-2021-0537

[42] Grover S, Wen W, Viswanathan V, Gill CT, Reinhart RMG. Long-lasting, dissociable improvements in working memory and long-term memory in older adults with repetitive neuromodulation. *Nat Neurosci*. 2022;25(9):1237-1246. doi:10.1038/s41593-022-01132-3

[43] Shapira R, Gdalyahu A, Gottfried I, et al. Hyperbaric oxygen therapy alleviates vascular dysfunction and amyloid burden in an Alzheimer's disease mouse model and in elderly patients. *Aging (Albany NY)*. 2021;13(17):20935-20961. doi:10.18632/aging.203485

[44] Chen J, Zhang F, Zhao L, et al. Hyperbaric oxygen ameliorates cognitive impairment in patients with Alzheimer's disease and amnestic mild cognitive impairment. *Alzheimers Dement (N Y)*. 2020;6(1):e12030. doi:10.1002/trc2.12030

[45] Riazifar M, Mohammadi MR, Pone EJ, et al. Stem cell-derived exosomes as nanotherapeutics for autoimmune and neurodegenerative disorders. *ACS Nano*. 2019;13(6):6670-6688. doi:10.1021/acsnano.9b01004

[46] Kang HK, Liu M, Datta SK. Low-dose peptide tolerance therapy of lupus generates plasmacytoid dendritic cells that cause expansion of autoantigen-specific regulatory T cells and contraction of inflammatory Th17 cells. *J Immunol*. 2007;178(12):7849-7858. doi:10.4049/jimmunol.178.12.7849

[47] Lopez P, de Paz B, Rodriquez-Carrio J, et al. Th17 responses and natural IgM antibodies are related to gut microbiota composition in systemic lupus erythematosus patients. *Sci Rep*. 2016;6:24072. doi:10.1038/srep24072

Chapter 13

[1] Liang H, Tian X, Lan-Yu C, Yan-Yan C, Wang C. Effect of psychological intervention on health-related quality of life in people with systemic lupus erythematosus: a systematic review. *Int J Nurs Sci*. 2014;2(3):298-305. doi:10.1016/j.ijnss.2014.07.008

[2] Grabich S, Farrelly E, Ortmann R, et al. (2022). Real-world burden of systemic lupus erythematosus in the USA: a comparative cohort study from the Medical Expenditure Panel Survey (MEPS) 2016–2018. *Lupus Science & Medicine*; **9**:e000640. DOI: 10.1136/lupus-2021-000640

[3] Strand, V., Simon, L.S., Meara, A.S., & Touma, Z. (2020). Measurement properties of selected patient-reported outcome measures for use in randomized controlled trials in patients with systemic lupus erythematosus: a systematic review. *Lupus Science & Medicine,* 7(1). DOI: 10.1136/lupus-2019-000373

[4] Barraclough M, McKie S, Parker B, Elliott R, Bruce I. The effects of disease activity on neuronal and behavioural cognitive processes in systemic lupus erythematosus. *Rheumatology (Oxford).* Published online March 24, 2021:keab256. doi:10.1093/rheumatology/keab256

[5] Bendorius M, Po C, Jeltsch-David H. From systemic inflammation to neuroinflammation: the case of neurolupus. *Int J Mol Sci.* 2018;19(11):3588. doi:10.3390/ijms19113588

[6] Monahan R, Beaart-van de Voorde LJ, Steup-Beekman G, et al. Neuropsychiatric symptoms in systemic lupus erythematosus: impact on quality of life. *Lupus.* 2017;26(12):1252-1259. doi:10.1177/0961203317694262

[7] Olah C, Schwartz N, Denton C, et al. Cognitive dysfunction in autoimmune rheumatic diseases. *Arthritis Res Ther.* 2020;22:78. doi:10.1186/s13075-020-02180-5

[8] Young K, Sen D, Drummond R, et al. Cognitive impairment and participation in systemic lupus erythematosus. *Am J Occup Ther.* 2020;64(Suppl 1):7411505271p1. doi:10.5014/ajot.2020.74S1-P09721

[9] Khoury LE, Zarfeshani A, Diamond B. Using the mouse to model human diseases: cognitive impairment in systemic lupus erythematosus. *J Rheumatol.* 2020;47(7):1145-1149. doi:10.3899/jrheum.200410

[10] Aguirre E, Woods RT, Spector A, Orrell M. Cognitive stimulation for dementia: a systematic review of the evidence of effectiveness from randomized controlled trials. *Ageing Res Rev.* 2013;12(1):253-262. doi:10.1016/j.arr.2012.07.001

[11] Hepsomali P, Coxon C. Inflammation and diet: focus on mental and cognitive health. *Adv Clin Exp Med.* 2022;31(8):821-825. doi:10.17219/acem/152350

[12] Castro-Webb N, Cozier YC, Barbhaiya M, et al. Association of macronutrients and dietary patterns with risk of systemic lupus erythematosus in the Black Women's Health Study. *Am J Clin Nutr.* 2021;114(4):1486-1494. doi:10.1093/ajcn/nqab224

[13] Rossato S, Oakes EG, Barbhaiya M, et al. Ultraprocessed food intake and risk of systemic lupus erythematosus among women observed in the Nurses' Health Study cohorts. *Arthritis Care Res (Hoboken).* 2024;76(6):907-913. doi:10.1002/acr.25395

[14] Ceccarelli F, Perricone C, Pirone C, et al. Cognitive dysfunction improves in systemic lupus erythematosus: results of a 10-year prospective study. *PLoS One.* 2018;13(5):e0196103. doi:10.1371/journal.pone.0196103

[15] Govoni M, Bombardieri S, Bortoluzzi A, et al. Factors and comorbidities associated with first neuropsychiatric event in systemic lupus erythematosus: does a risk profile exist? A large multicenter retrospective cross-sectional study on 959 Italian patients. *Rheumatology.* 2012;51(1):157-168. doi:10.1093/rheumatology/ker310

[16] Masley SM. *The Better Brain Solution: How to Start Now at Any Age and Prevent Insulin Resistance of the Brain, Sharpen Cognitive Function, and Avoid Memory Loss.* New York: Alfred A. Knopf; 2018. Available from: https://a.co/d/9zg2Ixo

[17] Willis L, Shukitt-Hale B, Cheng V, Joseph J. Dose-dependent effects of walnuts on motor and cognitive function in aged rats. *Br J Nutr.* 2009;101(8):1140-1144. doi:10.1017/S0007114508059369

[18] Lei F, Cheah I, Ng MM, et al. The association between mushroom consumption and mild cognitive impairment: A community-based cross-sectional study in Singapore. *J Alzheimers Dis.* 2019;68(1):197-203. doi:10.3233/JAD-180959

[19] Kraman M, Capelija E, Raseta M, Rakic M. Diversity, chemistry, and environmental contamination of wild growing medicinal mushroom species as sources of biologically active substances. In: *Biology, Cultivation and Applications of Mushrooms*. March 2022. doi:10.1007/978-981-16-6257-7_8

[20] Martínez-Mármol R, Chai Y, Conroy JN, et al. Hericerin derivatives activate a pan-neurotrophic pathway in central hippocampal neurons converging to ERK1/2 signaling enhancing spatial memory. *J Neurochem*. 2023;165(6). doi:10.1111/jnc.15767

Chapter 14

[1] Touma A, Moghaddam B, Su J, Katz P. Cognitive function trajectories are associated with the depressive symptom's trajectories in systemic lupus erythematosus over time. *Arthritis Care Res (Hoboken)*. 2021;73(10):1405–1413. doi:10.1002/acr.24349

[2] Mak A, Ho R, Tng H, Koh H, Chong JSX, Zhou J. Early cerebral volume reductions and their associations with reduced lupus disease activity in patients with newly-diagnosed systemic lupus erythematosus. *Sci Rep*. 2016;6:22231. doi:10.1038/srep22231

[3] Ceccarelli F, Perricone C, Pirone C, et al. Cognitive dysfunction improves in systemic lupus erythematosus: results of a 10-year prospective study. *PLoS One*. 2018;13(5):e0196103. doi:10.1371/journal.pone.0196103

[4] Nantes S, Su J, Dhaliwal A, Colosimo K, Touma Z. Performance of screening tests for cognitive impairment in systemic lupus erythematosus. *J Rheumatol*. 2017;44(11):1583–1589. doi:10.3899/jrheum.161125

[5] Holliday SL, Navarrete MG, Hermosilio-Romo D, et al. Validating a computerized neuropsychological test battery for mixed ethnic lupus patients. *Lupus*. 2003;12:693–703. doi:10.1191/0961203303lu442oa

[6] Yue R, Gurung I, Long XX, Xian JY, Peng XB. Prevalence, involved domains, and predictor of cognitive dysfunction in systemic lupus

erythematosus. *Lupus*. 2020;29(13):1743-1751. doi:10.1177/0961203320958061

[7] Wallace DJ, Hahn BH. *Dubois' Lupus Erythematosus*. 7th ed. Lippincott Williams & Wilkins; 2007. ISBN: 978-0-7817-9394-0. Accessed April 18, 2025. https://www.wolterskluwer.com/en/solutions/ovid/dubois-lupus-erythematosus-3484

[8] Hanly J, Omisade A, Su L, Farewell V, Fisk JD. Cognitive function in systemic lupus erythematosus, rheumatoid arthritis and multiple sclerosis assessed by computerized neuropsychological tests. *Arthritis Rheumatol*. 2010;62(5):1478-1486. doi:10.1002/art.27404

[9] Carlomagno S, Migliaresi S, Ambrosone I, Sannino M, Sanges G, Di Iorio G. Cognitive impairment in systemic lupus erythematosus: a follow-up study. *J Neurol*. 2000;247:273-279. doi:10.1007/s004150050583

[10] Plantinga L, Lim S, Bowling C, Drenkard C. Perceived stress and reported cognitive symptoms among Georgia patients with systemic lupus erythematosus. *Lupus*. 2017;26(10):1064-1071. doi:10.1177/0961203317693095

[11] Griffin RM. Lupus fog and memory problems. WebMD. https://www.webmd.com/lupus/features/lupus-fog-memory-problems#1. Accessed April 26, 2025.

[12] Rayes HA, Tani C, Kwan A, et al. What is the prevalence of cognitive impairment in lupus and which instruments are used to measure it? A systematic review and meta-analysis. *Semin Arthritis Rheum*. 2018;48(2):240-255. doi:10.1016/j.semarthrit.2018.02.007

[13] Hong YJ, Jang EH, Hwang J, Roh JH, Lee JH. The efficacy of cognitive intervention programs for mild cognitive impairment: a systematic review. *Curr Alzheimer Res*. 2015;12(6). doi:10.2174/1567205012666150530201636

[14] Lapa A, Postal M, Sinicato N, et al. S100B is associated with cognitive impairment in childhood-onset systemic lupus erythematosus patients. *Lupus*. 2017;26:478-483. doi:10.1177/0961203317691374

[15] Magro-Checa C, Seup-Beekman G, Huizinga T, van Buchem MA, Ronen I. Laboratory and neuroimaging biomarkers in neuropsychiatric systemic lupus erythematosus: Where do we stand, where to go? *Front Med (Lausanne)*. 2018;5:340. doi:10.3389/fmed.2018.00340

[16] Benedict R, Shucard J, Zivadinov R, Shucard D. Neuropsychological impairment in systemic lupus erythematosus: a comparison with multiple sclerosis. *Neuropsychol Rev*. 2008;18(2):149-166. doi:10.1007/s11065-008-9061-2

[17] Conti F, Alessandri C, Perricone C, et al. Neurocognitive dysfunction in systemic lupus erythematosus: Association with antiphospholipid antibodies, disease activity, and chronic damage. *PLoS One*. 2012;7(3):e33824. doi:10.1371/journal.pone.0033824

[18] Appenzeller S, Rondina JM, Li LM, Costallat LT, Cendes F. Cerebral and corpus callosum atrophy in systemic lupus erythematosus. *Arthritis Rheum*. 2005;52(9):2783-2789. doi:10.1002/art.21271

[19] Kello N, Anderson E, Diamond B. Cognitive dysfunction in systemic lupus erythematosus: a case for initiating trials. *Arthritis Rheumatol*. 2019;71(9):1413-1425. doi:10.1002/art.40933

[20] Jeltsch-David H, Muller S. Autoimmunity, neuroinflammation, pathogen load: a decisive crosstalk in neuropsychiatric SLE. *J Autoimmun*. 2016;74:13-26. doi:10.1016/j.jaut.2016.04.005

[21] Barraclough M, McKie S, Parker B, et al. Altered cognitive function in systemic lupus erythematosus and associations with inflammation and functional and structural brain changes. *Ann Rheum Dis*. 2019;78:934-940. doi:10.1136/annrheumdis-2018-214677

[22] Barraclough M, McKie S, Parker B, Elliott R, Bruce I. The effects of disease activity on neuronal and behavioural cognitive processes in

systemic lupus erythematosus. *Rheumatology (Oxford)*. Published online March 24, 2021. doi:10.1093/rheumatology/keab256

[23] Lesher AI, Landis S, Stroud C, Downey A. *Preventing Cognitive Decline and Dementia: A Way Forward*. Washington, DC: The National Academies Press; 2017. doi:10.17226/24782

[24] Masley SM. *The Better Brain Solution: How to Start Now at Any Age and Prevent Insulin Resistance of the Brain, Sharpen Cognitive Function, and Avoid Memory Loss*. New York: Alfred A. Knopf; 2018. Available at: https://a.co/d/9zg2Ixo

[25] Manly JJ, Jones RN, Langa KM, et al. Estimating the prevalence of dementia and cognitive impairment in the US: The 2016 health and retirement study harmonized cognitive assessment protocol project. *JAMA Neurol*. 2022;79(12). doi:10.1001/jamaneurol.2022.3543

Chapter 15

[1] McGrattan A, McEvoy C, McGuinness B, McKinley MC, Woodside JV. Effect of dietary interventions in mild cognitive impairment: a systematic review. *Br J Nutr*. 2018;120(12):1388-1405. doi:10.1017/S0007114518002945

[2] Ngandu T, Lehtisalo J, Solomon A, et al. A 2-year multidomain intervention of diet, exercise, cognitive training, and vascular risk monitoring versus control to prevent cognitive decline in at-risk elderly people (FINGER): a randomised controlled trial. *Lancet*. 2015;385(9984):2255-2263. doi:10.1016/S0140-6736(15)60461-5

[3] Rouse HJ, Small BJ, Faust ME. Assessment of cognitive training & social interaction in people with mild to moderate dementia: a pilot study. *Clin Gerontol*. 2019;42(4):421-434. doi:10.1080/07317115.2019.1590489

[4] Bredesen DE. *The End of Alzheimer's Program: The First Protocol to Enhance Cognition and Reverse Decline at Any Age*. New York, NY:

Penguin Random House; 2020. https://a.co/d/5LwgXlt. Accessed April 26, 2025.

[5] Toups K, Hathaway A, Gordon D, et al. Precision medicine approach to Alzheimer's disease: successful proof-of-concept trial. *J Alzheimers Dis*. 2022;88(4):doi:10.3233/JAD-215707

[6] Masley SM. *The Better Brain Solution: How to Start Now at Any Age and Prevent Insulin Resistance of the Brain, Sharpen Cognitive Function, and Avoid Memory Loss*. New York, NY: Alfred A. Knopf; 2018.

[7] Thomas DE. *The Lupus Encyclopedia: A Comprehensive Guide for Patients and Families*. Baltimore, MD: A Johns Hopkins Press Health Book; 2014. Accessed April 24, 2025. https://a.co/d/8Vp2WES

[8] Charisis S, Ntanasi E, Yannakoulia M, et al. Diet inflammatory index and dementia incidence: a population-based study. *Neurology*. 2021;97(24). doi:10.1212/WNL.0000000000012973

[9] Haase S, Haghikia A, Wilck N, Müller DN, Linker RA. Impacts of microbiome metabolites on immune regulation and autoimmunity. *Immunology*. 2018;154(2):230-238. doi:10.1111/imm.12933

[10] Zhu S, Jiang Y, Xu K, et al. The progress of gut microbiome research related to brain disorders. *J Neuroinflammation*. 2020;17(1):25. doi:10.1186/s12974-020-1705-z

[11] Doidge N. *The Brain That Changes Itself: Stories of Personal Triumph from the Frontiers of Brain Science*. Penguin Books; 2007. https://a.co/d/e1dpLXv

[12] Aghjayan S, Bournias T, Kang C, et al. Aerobic exercise improves episodic memory in late adulthood: a systematic review and meta-analysis. *Commun Med*. 2022. doi:10.1038/s43856-022-00079-7

[13] O'Bryan T. *You Can Fix Your Brain*. New York, NY: Rodale Books, an imprint of Penguin Random House; 2018. https://a.co/d/hlbhEzA

[14] Fasano A. All disease begins in the (leaky) gut: role of zonulin-mediated gut permeability in the pathogenesis of some chronic inflammatory diseases. *F1000Res*. 2020;9:F1000 Faculty Rev-69. doi:10.12688/f1000research.20510.1

[15] Pall ML. Low intensity electromagnetic fields act via voltage-gated calcium channel (VGCC) activation to cause very early onset Alzheimer's disease: 18 distinct types of evidence. *Curr Alzheimer Res*. 2022. doi:10.2174/1567205019666220202114510

[16] Hyman M. *Food: What the Heck Should I Eat?* New York, NY: Little Brown & Co; 2018. https://a.co/d/bICyLEu

[17] Hyman M. *The Pegan Diet*. New York, NY: Little Brown Spark; 2021. https://a.co/d/cHk7Ly8

[18] Hyman M. *The Ultramind Solution*. Scribner Book Company; 2010. https://a.co/d/bIuEz93

[19] Ramos A, Garrison V, Koop D, et al. Polyphenon E, a green tea extract, increases brain NAA levels in MS: a pilot six-month open label study. *Neurology*. 2012;78(1). https://www.neurology.org/doi/10.1212/WNL.78.1_supplement.P03.050

[20] Deecken C. Can drinking tea improve cognitive function and help prevent cognitive decline? *Neuroscience News*. 2022. https://neurosciencenews.com/tes-cognition-20482/

[21] American Academy of Neurology. Adding color to your plate may lower risk of cognitive decline. *ScienceDaily*. Published July 29, 2021. Accessed November 10, 2021. https://www.sciencedaily.com/releases/2021/07/210729122215.htm

[22] Li W, Qin L, Feng R, et al. Emerging senolytic agents derived from natural products. *Mech Ageing Dev*. 2019;181:1-6. doi:10.1016/j.mad.2019.05.001

[23] Melo van Lent D, O'Donnell A, Beiser AS, et al. MIND diet adherence and cognitive performance in the Framingham Heart Study. *J Alzheimers Dis*. 2021;82(2):827-839. doi:10.3233/JAD-201238

[24] Rush University Medical Center. MIND diet linked to better cognitive performance: Study finds diet may contribute to cognitive resilience in the elderly. *ScienceDaily*. Published September 21, 2021. Accessed April 8, 2022. www.sciencedaily.com/releases/2021/09/210921172721.htm

[25] Dominguez LJ, Barbagallo M. Nutritional prevention of cognitive decline and dementia. *Acta Biomed*. 2018;89(2):276-290. doi:10.23750/abm.v89i2.7401

[26] Doria A. Targeting remission and low disease activity in SLE. *BMJ*. 2021;8(1). doi:10.1007/s40744-023-00601-w

[27] Mann D. Many people may be eating their way to dementia. *U.S. News*. Published November 11, 2021. Accessed November 10, 2021. https://www.usnews.com/news/health-news/articles/2021-11-11/many-people-may-be-eating-their-way-to-dementia

[28] Butler MJ, Deems NP, Muscat S, et al. Dietary DHA prevents cognitive impairment and inflammatory gene expression in aged male rats fed a diet enriched with refined carbohydrates. *Brain Behav Immun*. 2021;98:198-209

[29] Rossato S, Oakes EG, Barbhaiya M, et al. Ultraprocessed food intake and risk of systemic lupus erythematosus among women observed in the Nurses' Health Study cohorts. *Arthritis Care Res (Hoboken)*. Published online June 27, 2024. doi:10.1002/acr.25395

[30] Charisis S, Ntanasi E, Yannakoulia M, et al. Diet inflammatory index and dementia incidence: a population-based study. *Neurology*. 2021;97(24). doi:10.1212/WNL.0000000000012973

[31] Leiter O, Zhuo Z, Rust R, et al. Selenium mediates exercise-induced adult neurogenesis and reverses learning deficits induced by hippocampal injury and aging. *Cell Metab*. 2022. doi:10.1016/j.cmet.2022.01.005

[32] Meltzer A, Rose MK, Le AY, et al. Improvement in executive function for older adults through smartphone apps: a randomized clinical trial comparing language learning and brain training. *Aging Neuropsychol Cogn.* 2021. doi:10.1080/13825585.2021.1991262

[33] Stine-Morrow EAL, McCall GS, Manavbasi I, Ng S, Llano DA, Barbey AK. The effects of sustained literacy engagement on cognition and sentence processing among older adults. *Front Psychol.* 2022;13:923795. doi:10.3389/fpsyg.2022.923795

[34] Scharre DQ, Chang S, Nagaraja HN, Vrettos NE, Bornstein RA. Digitally translated self-administered Gerocognitive Examination (eSAGE): relationship with its validated paper version, neuropsychological evaluations, and clinical assessment. *Alzheimers Res Ther.* 2017;9(1):44. doi:10.1186/s13195-017-0269-3

[35] Peterson RC, Lopez O, Armstrong MJ, et al. Practice guideline update: mild cognitive impairment. Report of the guideline development, dissemination, and implementation subcommittee of the American Academy of Neurology. *Neurology.* 2017;88(10):1-10. doi:10.1212/WNL.0000000000004826

[36] El-Sayes J, Harasym D, Turco CV, Locke MB, Nelson AJ. Exercise-induced neuroplasticity: a mechanistic model and prospects for promoting plasticity. *Neuroscientist.* 2019;25(1):65-85. doi:10.1177/1073858418771538

[37] Gibbons TD, Cotter JD, Ainslie PN, et al. Fasting for 20 h does not affect exercise-induced increases in circulating BDNF in humans. *J Physiol.* Published online 2023. doi:10.1113/JP283582

[38] Gorzelitz J, Trabert B, Katki HA, et al. Independent and joint associations of weightlifting and aerobic activity with all-cause, cardiovascular disease and cancer mortality in the Prostate, Lung, Colorectal and Ovarian Cancer Screening Trial. *Br J Sports Med.* Published online September 27, 2022. doi:10.1136/bjsports-2021-105315

[39] Muiños M, Ballesteros S. Does dance counteract age-related cognitive and brain declines in middle-aged and older adults? A systematic review.

Neurosci Biobehav Rev. 2021;121:259-276. doi:10.1016/j.neubiorev.2020.11.028

[40] Cui L, Tao S, Yin HC, et al. Tai Chi Chuan alters brain functional network plasticity and promotes cognitive flexibility. *Front Psychol.* 2021;12:665419. doi:10.3389/fpsyg.2021.665419

[41] Gard T, Holzel BK, Lazar SW. The potential effects of meditation on age-related cognitive decline: a systematic review. *Ann N Y Acad Sci.* 2014;1307:89-103. doi:10.1111/nyas.12348

[42] Menigoz W, Latz TT, Ely RA, Kamei C, Melvin G, Sinatra D. Integrative and lifestyle medicine strategies should include earthing (grounding): review of research evidence and clinical observations. *Explore (NY).* 2020;16(3):152-160. doi:10.1016/j.explore.2019.10.005

[43] Ober C, Sinatra ST, Zucker M. *Earthing: The Most Important Health Discovery Ever!* 2nd ed. Basic Health Publications, Inc.; 2014. https://a.co/d/aPCxhiW

[44] Jimenez MP, Elliott EG, DeVille NV, et al. Residential green space and cognitive function in a large cohort of middle-aged women. *JAMA Netw Open.* 2022;5(4):e229306. doi:10.1001/jamanetworkopen.2022.9306

[45] Pandey KB, Rizvi SI. Plant polyphenols as dietary antioxidants in human health and disease. *Oxid Med Cell Longev.* 2009;2(5):270-278. doi:10.4161/oxim.2.5.9498

[46] Krikorian R, Skelton MR, Summer SS, Shidler MD, Sullivan PG. Blueberry supplementation in midlife for dementia risk reduction. *Nutrients.* 2022;14(8). doi:10.3390/nu14081619

[47] Samani P, Costa S, Cai S. Neuroprotective effects of blueberries through inhibition on cholinesterase, tyrosinase, cyclooxygenase-2, and amyloidogenesis. *Nutraceuticals.* 2023;3(1):39-57. doi:10.3390/nutraceuticals3010004

[48] Chauhan A, Chauhan V. Beneficial effects of walnuts on cognition and brain health. *Nutrients*. 2020;12(2):550. doi:10.3390/nu12020550

[49] Nakhaee S, Kooshki A, Hormozi A, et al. Cinnamon and cognitive function: a systematic review of preclinical and clinical studies. *Nutr Neurosci*. Published online 2023. doi:10.1080/1028415X.2023.2166436

[50] M. Lupus brain fog: A biologic perspective on cognitive impairment, depression, and fatigue in systemic lupus erythematosus. *Immunol Res*. 2015;63:26-37. doi:10.1007/s12026-015-8716-3Mackay, M.

[51] Alzheimer's Disease International. *World Alzheimer Report 2021: Journey Through the Diagnosis of Dementia*. London, UK: Alzheimer's Disease International; 2021. https://www.alzint.org/resource/world-alzheimer-report-2021/

[52] Plantinga L, Vandenberg A, Goldstein F, et al. Patient and provider perceptions of a novel cognitive functioning report for patients with systemic lupus erythematosus: a qualitative study. *Lupus Sci Med*. 2021;8:e000476. doi:10.1136/lupus-2021-000476

Chapter 16

[1] Chandler MJ, Parks AC, Marsiske M, Rotblatt LJ, Smith GE. Everyday impact of cognitive interventions in mild cognitive impairment: a systematic review and meta-analysis. *Neuropsychol Rev*. 2017;26(3):225-251. doi:10.1007/s11065-016-9330-4

[2] Clare L, Woods RT, Moniz Cook ED, Orrell M, Spector A. Cognitive rehabilitation and cognitive training for early-stage Alzheimer's disease and vascular dementia. *Cochrane Database Syst Rev*. 2003;(4):CD003260. doi:10.1002/14651858.CD003260

[3] Monahan R, Beaart-van de Voorde LJ, Steup-Beekman G, et al. Neuropsychiatric symptoms in systemic lupus erythematosus: impact on quality of life. *Lupus*. 2017;26(12):1252-1259. doi:10.1177/0961203317694262

[4] Barrios PG, Pabon RG, Hanna SM, et al. Priority of treatment outcomes for caregivers and patients with mild cognitive impairment: preliminary analyses. *Neurol Ther.* 2016;5(1):1-10. doi:10.1007/s40120-016-0049-1

[5] Fleck C, Corwin M. Evidence-based decisions: memory intervention for individuals with mild cognitive impairment. *EBP Briefs.* 2013;8:1-14. https://www.pearsonassessments.com/content/dam/school/global/clinical/us/assets/ebp-briefs/EBPV8A4.pdf

[6] Manzine PR, Pavarini SCI. Cognitive rehabilitation: literature review based on levels of evidence. *Dement Neuropsychol.* 2009;3(3):248-255. doi:10.1590/S1980-57642009DN30300012

[7] Howick J, Chalmers I, Glasziou P, et al. Explanation of the 2011 Oxford CEBM Levels of Evidence (Introductory Document). *Oxford Centre for Evidence-Based Medicine.* 2011. https://www.cebm.net/index.aspx?o=5653

[8] Tamilou F, Arnaud L, Talarico R, et al. Systemic lupus erythematosus: state of the art of clinical practice guidelines. *RMD Open.* 2018;4(1):e000793. doi:10.1136/rmdopen-2018-000793

[9] Devanand DP, Goldberg TE, Qian M, et al. Computerized games versus crosswords training in mild cognitive impairment. *NEJM Evid.* 2022;1(12). doi:10.1056/evidoa2200121

[10] Edwards J, Fausto B, Tetlow A, Corona RT, Valdes EG. Systematic review and meta-analyses of useful field of view cognitive training. *Neurosci Biobehav Rev.* 2018;84:72–91. doi:10.1016/j.neubiorev.2017.11.004

[11] Langenbah DM, Ashman T, Cantor J, Trott C. An evidence-based review of cognitive rehabilitation in medical conditions affecting cognitive function. *Arch Phys Med Rehabil.* 2013;94:271–286. doi:10.1016/j.apmr.2012.09.011

[12] Harrison MJ, Morris KA, Horton R, et al. Results of intervention for lupus patients with self-perceived cognitive difficulties. *Neurology.* 2005;65(8):1325–1327. doi:10.1212/01.wnl.0000180938.69146.5e

[13] Mason-Baughman M, Raupp S, Corman K. Staging and treatment frameworks for dementia management. *Perspect ASHA Spec Interest Groups*. 2016;1(1). doi:10.1044/persp1.SIG15.53

[14] Murray LL, Paek EJ. Behavioral/nonpharmacological approaches to addressing cognitive-linguistic symptoms in individuals with dementia. *Perspect ASHA Spec Interest Groups*. 2016;1(Sig 15). doi:10.1044/persp1.SIG15.12

[15] Benigas JE, Brush JA, Elliot GM. *Spaced Retrieval: Step by Step*. Baltimore, MD: Health Professions Press; 2016. https://www.healthpropress.com/product/spaced-retrieval-step-by-step

[16] Sohlberg MM, Mateer CA. *Cognitive Rehabilitation: An Integrative Neuropsychological Approach*. New York, NY: The Guilford Press; 2001. https://a.co/d/f2f6zOk

[17] Sohlberg MM, Mateer CA. *Introduction to Cognitive Rehabilitation: Theory & Practice*. New York, NY: The Guilford Press; 1989. https://a.co/d/0kGgJBl

[18] Edwards JD, Xu H, Clark DO, Guey LT, Ros LA, Unverzagt FW. Speed of processing training results in lower risk of dementia. *Alzheimers Dement (Amst)*. 2017;3(4):603–611. doi:10.1016/j.trci.2017.09.002

[19] Barutca CD, Gokce S. Influence of chronic disease on cognitive functions of patients. *Int J Caring Sci*. 2020;13(1):315. https://www.internationaljournalofcaringsciences.org/docs/36_%20barutcu_original_13_1_2.pdf

[20] Scharre DQ, Chang S, Nagaraja HN, Wheeler NC, Kataki M. Self-Administered Gerocognitive Examination: longitudinal cohort testing for the early detection of dementia conversion. *Alzheimers Res Ther*. 2021;13:103. doi:10.1186/s13195-021-00829-7

[21] Alzheimer's Disease International. *World Alzheimer Report 2021: Journey Through the Diagnosis of Dementia*. London, UK: Alzheimer's

Disease International; 2021. https://www.alzint.org/resource/world-alzheimer-report-2021/

[22] Sood A, Al H, Raji M. Cognitive impairment in elderly patients with rheumatic disease and the effect of disease-modifying anti-rheumatic drugs. *Clin Rheumatol*. 2021. doi:10.1007/s10067-020-05372-1

[23] Plantinga L, Vandenberg A, Goldstein F, et al. Patient and provider perceptions of a novel cognitive functioning report for patients with systemic lupus erythematosus: a qualitative study. *Lupus Sci Med*. 2021;8:e000476. doi:10.1136/lupus-2021-000476

[24] Fleming VB, Harris JL. Toward identifying mild cognitive impairment in Hispanic and African-American adults. *Perspect ASHA Spec Interest Groups*. 2017;2(3):110–118. doi:10.1044/persp2.SIG2.110

[25] Kello N, Anderson E, Diamond B. Cognitive dysfunction in systemic lupus erythematosus: A case for initiating trials. *Arthritis Rheumatol*. 2019;71(9):1413–1425. doi:10.1002/art.40933

[26] Barraclough M, McKie S, Parker B, et al. Altered cognitive function in systemic lupus erythematosus and associations with inflammation and functional and structural brain changes. *Ann Rheum Dis*. 2019;78:934–940. doi:10.1136/annrheumdis-2018-214677

[27] Yue R, Gurung I, Long XX, Xian JY, Peng XB. Prevalence, involved domains, and predictor of cognitive dysfunction in systemic lupus erythematosus. *Lupus*. 2020;29(13):1743–1751. doi:10.1177/0961203320958061

[28] Geddes MR, O'Connell ME, Fisk JD, et al. Remote cognitive and behavioral assessment: Report of the Alzheimer Society of Canada Task Force on dementia care best practices for COVID-19. *Alzheimers Dement (Amst)*. 2020;12(1):e12111. doi:10.1002/dad2.12111

[29] Raghunath S, Glikmann-Johnston Y, Morand E, Stout JC, Hoi A. Evaluation of the Montreal Cognitive Assessment as a screening tool for cognitive dysfunction in SLE. *Lupus Sci Med*. 2021;8:e000580. doi:10.1136/lupus-2021-000580

[30] Barutca CD, Gokce S. Influence of chronic disease on cognitive functions of patients. *Int J Caring Sci*. 2020;13(1):315.

[31] He JW, Diaz Martinez JP, Bingham K, Su J, et al. Insight into intraindividual variability across neuropsychological tests and its association with cognitive dysfunction in patients with lupus. *Lupus Sci Med*. 2021;8(1):e000511. doi:10.1136/lupus-2021-000511

[32] Fleming VB, Harris JL. Toward identifying mild cognitive impairment in Hispanic and African-American adults. *Perspect ASHA Spec Interest Groups*. 2017;2(3). doi:10.1044/persp2.SIG2.110

[33] Rayes HA, Tani C, Kwan A, et al. What is the prevalence of cognitive impairment in lupus and which instruments are used to measure it? A systematic review and meta-analysis. *Semin Arthritis Rheum*. 2018;48(2):240-255. doi:10.1016/j.semarthrit.2018.02.007

[34] Fernandez H, Cevallos A, Sotomayor RJ, et al. Mental disorders in systemic lupus erythematosus: a cohort study. *Rheumatol Int*. 2019;39. doi:10.1007/s00296-019-04423-4

[35] El-Shafey AM, Abd-El-Geleel SM, Soliman ES. Cognitive impairment in non-neuropsychiatric systemic lupus erythematosus. *Egypt Rheumatol*. 2012;34:67-73. doi:10.1016/j.ejr.2012.02.002

[36] Hanly JG, Legge A, Kamintsky L, et al. Role of autoantibodies and blood-brain barrier leakage in cognitive impairment in systemic lupus erythematosus. *Lupus Sci Med*. 2022;9(1):e000668. doi:10.1136/lupus-2022-000668

[37] Haskins EC. *Cognitive Rehabilitation Manual: Translating Evidence-Based Recommendations into Practice*. American Congress of Rehabilitation Medicine; 2012. https://a.co/d/3zoH9aP

[38] Sohlberg MM, Turkstra LS. *Optimizing Cognitive Rehabilitation: Effective Instructional Methods*. The Guilford Press; 2011. https://a.co/d/5VtdIOZ

[39] Cao ZY, Wang N, Jia JT, et al. Abnormal topological organization in systemic lupus erythematosus: a resting-state functional magnetic

resonance imaging analysis. *Brain Imaging Behav.* 2020. doi:10.1007/s11682-019-00228-y

[40] Leslie B, Crowe SF. Cognitive functioning in systemic lupus erythematosus: a meta-analysis. *Lupus.* 2018;27(6):920-929. doi:10.1177/0961203317751859

[41] Bogaczewicz J, Sysa-Jedrzejowska A, Arkuszewska C, et al. Vitamin D status in systemic lupus erythematosus patients and its association with selected clinical and laboratory parameters. *Lupus.* 2012;21(5):477-484. doi:10.1177/0961203311427549

[42] Ceccarelli F, Pirone C, Mina C, et al. Pragmatic language dysfunction in systemic lupus erythematosus patients: Results from a single center Italian study. *PLoS One.* 2019;14(11):e0224437. doi:10.1371/journal.pone.0224437

[43] Prutting C, Kirchner D. A clinical appraisal of the pragmatic aspects of language. *J Speech Hear Disord.* 1987;52:105-119. doi:10.1044/jshd.5202.105

[44] Arcara G, Bambini V. A test for the assessment of pragmatic abilities and cognitive substrates (APACS): Normative data and psychometric properties. *Front Psychol.* 2016;7:70. doi:10.3389/fpsyg.2016.00070

[45] Rad DS. A review on adult pragmatic assessments. *Iran J Neurol.* 2014;13(3):113-118. PMID:25422728

[46] Belleville S, Cloutier S, Mellah S, et al. When is more better? Modeling the effect of dose on the efficacy of the MAPT multidomain intervention as a function of individual characteristics. *Alzheimers Dement.* 2021. doi:10.1002/alz.054948

[47] De Ville DV, Farouj Y, Preti MG, Liegeois R, Amico E. When makes you unique: Temporality of the human brain fingerprint. *Sci Adv.* 2021;7(42):eabj0751. doi:10.1126/sciadv.abj0751

[48] Rosenberg A, Mangialasche F, Ngandu T, Solomon A, Kivipelto M. Multidomain interventions to prevent cognitive impairment, Alzheimer's

disease, and dementia: From FINGER to World-Wide FINGERS. *J Prev Alzheimers Dis*. 2020;7(1):29-36. doi:10.14283/jpad.2019.41

[49] Bredesen DE. *The End of Alzheimer's Program: The First Protocol to Enhance Cognition and Reverse Decline at Any Age*. New York: Penguin Random House; 2020. https://a.co/d/5LwgXlt

Chapter 17

[1] Kozora E, Arciniegas DB, Duggan E, West S, Brown MS, Filley CM. White matter abnormalities and working memory impairment in systemic lupus erythematosus. *Cogn Behav Neurol*. 2013;26(2):63-72. doi:10.1097/WNN.0b013e31829d5c74

[2] Lampner C. Managing cognitive dysfunction in systemic lupus erythematosus. *Rheumatology Advisor*. Published 2018. Accessed April 24, 2025. https://www.rheumatologyadvisor.com/features/managing-cognitive-dysfunction-in-systemic-lupus-erythematosus/

[3] Rayes HA, Tani C, Kwan A, et al. What is the prevalence of cognitive impairment in lupus and which instruments are used to measure it? A systematic review and meta-analysis. *Semin Arthritis Rheum*. 2018;48(2):240-255. doi:10.1016/j.semarthrit.2018.02.007

[4] Cabeca H, Rocha L, Sabba A, et al. The subtleties of cognitive decline in multiple sclerosis: an exploratory study using hierarchical cluster analysis of CANTAB results. *BMC Neurol*. 2018;18:140. doi:10.1186/s12883-018-1141-1

[5] Özakbaş S. Cognitive Impairment in Multiple Sclerosis: Historical Aspects, Current Status, and Beyond. *Noro Psikiyatr Ars*. 2015;52(Suppl 1):S12-S15. doi:10.5152/npa.2015.12610

[6] Bassi MS, Garofalo S, Marfia GA, et al. Amyloid-β homeostasis bridges inflammation, synaptic plasticity deficits and cognitive dysfunction in multiple sclerosis. *Front Mol Neurosci*. 2017;10:390. doi:10.3389/fnmol.2017.00390

[7] Pierson S, Griffith N. Treatment of cognitive impairment in multiple sclerosis. *Behav Neurol.* 2006;17(1):53-67. doi:10.1155/2006/545860

[8] Langdon D, Amato M, Boringa J, et al. Recommendations for a brief international cognitive assessment for multiple sclerosis (BICAMS). *Mult Scler.* 2012;18(6):891-898. doi:10.1177/1352458511431076

[9] Migliore S, Ghazaryan A, Simonelli I, et al. Cognitive impairment in relapsing-remitting multiple sclerosis patients with very mild clinical disability. *Behav Neurol.* 2017;7404289. doi:10.1155/2017/7404289

[10] Legenfelder J, Bryant D, Diamond BJ, et al. Processing speed interacts with working memory efficiency in multiple sclerosis. *Arch Clin Neuropsychol.* 2006;21:229-238. doi:10.1016/j.acn.2005.12.001

[11] Matias-Guiu J, Cortes-Martinez A, Valles-Salgado M, et al. Functional components of cognitive impairment in multiple sclerosis: A cross-sectional investigation. *Front Neurol.* 2017;8:643. doi:10.3389/fneur.2017.00643

[12] Shucard JL, Hamlin AS, Shucard DW. The relationship between processing speed and working memory demand in Systemic Lupus Erythematosus: Evidence from a Visual N-Back Test. *Neuropsychology.* 2011;25(1):45-52. doi:10.1037/a0021218

[13] Ekmekci O. Pediatric multiple sclerosis and cognition: A review of clinical, neuropsychologic, and neuroradiologic features. *Behav Neurol.* 2017;1463570. doi:10.1155/2017/1463570

[14] Rimkus C, Steenwijk M, Barkhof F. Causes, effects and connectivity changes in MS-related cognitive decline. *Dement Neuropsychol.* 2016;10(1):2-11. doi:10.1590/S1980-57642016DN10100002

[15] MS, Garofalo S, Marfia GA, et al. Amyloid-B homeostasis bridges inflammation, synaptic plasticity deficits and cognitive dysfunction in multiple sclerosis. *Front Mol Neurosci.* 2017;10:390. doi:10.3389/fnmol.2017.00390Bassi, M.

[16] Matias-Guiu J, Cortes-Martinez A, Valles-Salgado M, et al. Functional components of cognitive impairment in multiple sclerosis: A cross-

sectional investigation. *Front Neurol.* 2017;8:643. doi:10.3389/fneur.2017.00643

[17] Musella A, Gentile A, Rizzo F, et al. Interplay between age and neuroinflammation in multiple sclerosis: Effects on motor and cognitive functions. *Front Aging Neurosci.* 2018;10:238. doi:10.3389/fnagi.2018.00238

[18] Calabrese M, Favaretto A, Martini V, Gallo P. Grey matter lesions in MS: from histology to clinical implications. *Prion.* 2013;7(1):20-27. doi:10.4161/pri.22580

[19] Laman J, Hart B, Power C, Dziarski R. Bacterial peptidoglycan as a driver of chronic brain inflammation. *Trends Mol Med.* 2020;26(7). doi:10.1016/j.molmed.2019.11.006

[20] Langeskov-Christensen M, Eskildsen S, Stenager E, et al. Aerobic capacity is not associated with most cognitive domains in patients with multiple sclerosis – a cross-sectional investigation. *J Clin Med.* 2018;7(9):272. doi:10.3390/jcm7090272

[21] Yang F, Wen PS, Behoux F, Zhao Y. Effects of vibration training on cognition and quality of life in people with multiple sclerosis. *J MS Care.* 2021. https://doi.org/10.7224/1537-2073.2020-095

[22] Langdon D, Amato M, Boringa J, et al. Recommendations for a Brief International Cognitive Assessment for Multiple Sclerosis (BICAMS). *Mult Scler.* 2012;18(6):891-898. doi:10.1177/1352458511431076

[23] Vanotti S, Caceres F. Cognitive and neuropsychiatric disorders among MS patients from Latin America. *Mult Scler J Exp Transl Clin.* 2017;3(3). doi:10.1177/2055217317717508

[24] Brochet B, Ruet A. Cognitive impairment in multiple sclerosis with regards to disease duration and clinical phenotypes. *Front Neurol.* 2019;10:261. doi:10.3389/fneur.2019.00261

[25] Rimkus C, Steenwijk M, Barkhof F. Causes, effects and connectivity changes in MS-related cognitive decline. *Dement Neuropsychol.* 2016;10(1):2-11. doi:10.1590/S1980-57642016DN10100002

[26] Ozakbas S, Yigit P, Cinar B, et al. The Turkish validation of the Brief International Cognitive Assessment for Multiple Sclerosis (BICAMS) battery. *BMC Neurol.* 2017;17:208. doi:10.1186/s12883-017-0993-0

[27] Benedict RH, Shucard J, Zivadinov R, Shucard D. Neuropsychological impairment in systemic lupus erythematosus: A comparison with multiple sclerosis. *Neuropsychol Rev.* 2008;18(2):149-166. doi:10.1007/s11065-008-9061-2

[28] Wallace DJ, Hahn BH. *Dubois' Lupus Erythematosus*. 7th ed. Philadelphia, PA: Lippincott Williams & Wilkins; 2007. ISBN: 978-0-78-179394-0

[29] Yuen K, Bingham K, Tayer-Shifman Ol, Touma Z. Measures of cognition in rheumatic disease. *Arthritis Care Res.* 2020;72(S10):660-675. doi:10.1002/acr.24364

[30] Boyd R, Bennett S, Mori T, et al. GM-CSF upregulated in rheumatoid arthritis reverses cognitive impairment and amyloidosis in Alzheimer mice. *J Alzheimers Dis.* 2010;21(2):507-518. doi:10.3233/JAD-2010-091471

[31] Policicchio S, Ahmad A, Powell J, Proitsi P. Rheumatoid arthritis and risk for Alzheimer's disease: a systematic review and meta-analysis and a Mendelian randomization study. *Sci Rep.* 2017;7:12861. doi:10.1038/s41598-017-13168-8

[32] Konig MF. The microbiome in autoimmune rheumatic disease. *Best Pract Res Clin Rheumatol.* 2020;34:101473. doi:10.1016/j.berh.2019.101473

[33] Van Vollenhoven RV. 10 Lessons from RA: RA drug development has advanced at a more rapid pace than SLE: What can we learn from our colleagues? *Lupus Sci Med.* 2021;8. doi:10.1136/lupus-2021-la.10

[34] Hahn J, Malspeis S, Choic MY, et al. Association of healthy lifestyle behaviors and the risk of developing rheumatoid arthritis among women. *Arthritis Care Res.* doi:10.1002/acr.24862

[35] Baptista T, Petersen L, Molina J, et al. Autoantibodies against myelin sheath and S100B are associated with cognitive dysfunction in patients with rheumatoid arthritis. *Clin Rheumatol.* 2017;36:1959-1968. doi:10.1007/s10067-017-3724-4

[36] Lapa A, Postal M, Sinicato N, et al. S100B is associated with cognitive impairment in childhood-onset systemic lupus erythematosus patients. *Lupus.* 2017;26:478-483. doi:10.1177/0961203317691374

[37] Andersson KM, Wasen C, Juzokaite L, et al. Inflammation in the hippocampus affects IGF1 receptor signaling and contributes to neurological sequelae in rheumatoid arthritis. *Proc Natl Acad Sci U S A.* 2018;115(51):E12063-E12072. doi:10.1073/pnas.1810553115

[38] Olah C, Schwartz N, Denton C, et al. Cognitive dysfunction in autoimmune rheumatic diseases. *Arthritis Res Ther.* 2020;22:78. doi:10.1186/s13075-020-02180-5

[39] Sood A, Raji M. Cognitive impairment in elderly patients with rheumatic disease and the effect of disease-modifying anti-rheumatic drugs. *Clin Rheumatol.* 2021. doi:10.1007/s10067-020-05372-1

[40] Baker NA, Barvour KE, Halmick CG, Zack M, Snih SA. Arthritis and cognitive impairment in older adults. *Rheumatol Int.* 2017;37(6):955-961. doi:10.1007/s00296-017-3698-1

[41] Veeranki SP, Downer B, Jupiter D, Wong R. Arthritis and risk of cognitive and functional impairment in older Mexican adults. *J Aging Health.* 2017;29(3):454-473. doi:10.1177/0898264316636838

[42] Julian LJ, Yazdany J, Trupin L, Criswell LA, Yelin E, Katz PP. Validity of brief screening tools for cognitive impairment in rheumatoid arthritis and systemic lupus erythematosus. *Arthritis Care Res.* 2012;64(3):448-454. doi:10.1002/acr.21566

[43] Shin SY, Katz PK, Julian L. The relationship between perceived cognitive dysfunction and objective neuropsychological performance in persons with rheumatoid arthritis. *Arthritis Care Res*. 2014;65(3):481-486. doi:10.1002/acr.21814

[44] Shin SY, Katz P, Wallhagen M, Julian L. Cognitive impairment in persons with rheumatoid arthritis. *Arthritis Care Res*. 2012;64(8):1144-1150. doi:10.1002/acr.21683

[45] Rattery G, He J, Pearce R, et al. Disease activity and cognition in rheumatoid arthritis: an open label pilot study. *Arthritis Res Ther*. 2012;14(6):R263. doi:10.1186/ar4108

[46] Boyd R, Bennett S, Mori T, et al. GM-CSF upregulated in rheumatoid arthritis reverses cognitive impairment and amyloidosis in Alzheimer mice. *J Alzheimers Dis*. 2010;21(2):507-518. doi:10.3233/JAD-2010-091471

[47] McGuinness B. Understanding whether drugs for rheumatoid arthritis can reduce the risk of Alzheimer's disease. Alzheimer's Society of the UK. Published March 2019. Accessed April 2025. https://dementiamap.uk/projects/understanding-whether-drugs-for-rheumatoid-arthritis-can-reduce-the-risk-of-alzheimers-disease/

[48] Shin SY, Katz P, Wallhagen M, Julian L. Cognitive impairment in persons with rheumatoid arthritis. *Arthritis Care Res*. 2012;64(8):1144-1150. doi:10.1002/acr.21683

[49] Hanly J, Omisade A, Su L, Farewell V, Fisk JD. Cognitive function in systemic lupus erythematosus, rheumatoid arthritis, and multiple sclerosis assessed by computerized neuropsychological tests. *Arthritis Rheumatol*. 2010;62(5):1478-1486. doi:10.1002/art.27404

[50] Rattery G, He J, Pearce R, et al. Disease activity and cognition in rheumatoid arthritis: an open label pilot study. *Arthritis Res Ther*. 2012;14(6):R263. doi:10.1186/ar4108

[51] Pavlakis P. Rheumatologic disorders and the nervous system. *Neurol Clin*. 2020;26(3):591-610. doi:10.1212/CON.0000000000000856

[52] Wiseman SJ, Bastin ME, Hamilton IF, et al. Fatigue and cognitive function in systemic lupus erythematosus: associations with white matter microstructural damage. A diffusion tensor MRI study and meta-analysis. *Lupus*. 2017;26:588-597. doi:10.1177/0961203316668417

[53] Pavlakis P. Rheumatologic disorders and the nervous system. *Neurol Clin*. 2020;26(3):591-610. doi:10.1212/CON.0000000000000856

[54] Costa AC, Nunes DPF, Julie PR, et al. Neuropsychiatric manifestations in systemic lupus erythematosus and Sjogren's disease. *Autoimmun Rev*. 2025;24(4). doi:10.1016/j.autrev.2025.103756

[55] Carlo MD, Becciolini A, Incorvaia A, et al. Mild cognitive impairment in psoriatic arthritis: prevalence and associated factors. *Medicine (Baltimore)*. 2021;100(11):e24833. doi:10.1097/MD.0000000000024833

[56] Chen KT, Chen YC, Fan YH, et al. Rheumatic diseases are associated with a higher risk of dementia: a nation-wide, population-based, case-control study. *Int J Rheum Dis*. 2018;21(2):373-380. doi:10.1111/1756-185X.13246

[57] Spain E. Putting lupus in permanent remission. Northwestern University. Published November 10, 2013. Accessed November 10, 2021. https://news.feinberg.northwestern.edu/2013/11/lupusremission/

[58] Fasano A. Zonulin, regulation of tight junctions, and autoimmune diseases. *Ann N Y Acad Sci*. 2012;1258(1):25-33. doi:10.1111/j.1749-6632.2012.06538.x

[59] Attree EA, Dancey CP, Keeling D, Wilson C. Cognitive function in people with chronic illness: inflammatory bowel disease and irritable bowel syndrome. *Appl Neuropsychol*. 2003;10:96-104. doi:10.1207/S15324826AN1002_05

[60] Li Z, Zhang Z, Zhang Z, Wang Z, Li H. Cognitive impairment after long COVID-19: current evidence and perspectives. *Front Neurol*. 2023;14:1239182. doi:10.3389/fneur.2023.1239182

[61] Zhao S, Martin EM, Reuken PA, et al. Long COVID is associated with severe cognitive slowing: a multicentre cross-sectional study. *eClinicalMedicine*. 2024;68:102434. doi:10.1016/j.eclinm.2024.102434

[62] Wood GK, Sargent BF, Ahmad ZUA, et al. Posthospitalization COVID-19 cognitive deficits at 1 year are global and associated with elevated brain injury markers and gray matter volume reduction. *Nat Med*. 2025;31:245-257. doi:10.1038/s41591-024-03309-8

[63] Kim MS, Lee H, Lee SW, et al. Long-term autoimmune inflammatory rheumatic outcomes of COVID-19: a binational cohort study. *Ann Intern Med*. 2024;177(3):291-302. doi:10.7326/M23-1831

[64] Heo Y-W, Jeon JJ, Ha MC, et al. Long-term risk of autoimmune and autoinflammatory connective tissue disorders following COVID-19. *JAMA Dermatol*. 2024;160(12). doi:10.1001/jamadermatol.2024.4233

[65] Sasa N, Kojima S, Koide R, et al. Blood DNA virome associates with autoimmune diseases and COVID-19. *Nat Genet*. 2025;57(1):65-79. doi:10.1038/s41588-024-02022-z

[66] Wang Y, Xie X, Zhang C, et al. Rheumatoid arthritis, systemic lupus erythematosus and primary Sjögren's syndrome shared megakaryocyte expansion in peripheral blood. *Ann Rheum Dis*. 2022;81(3). doi:10.1136/annrheumdis-2021-220066

[67] Pavlakis P. Rheumatologic disorders and the nervous system. *Neurol Clin Pract*. 2020;26(3):591-610. doi:10.1212/CON.0000000000000856

[68] Myasoedova E, et al. Cognitive impairment in individuals with rheumatic diseases: the role of systemic inflammation, immunomodulatory medications, and comorbidities. *Lancet Rheumatol*. 2024;6(12):e871-e880. doi:10.1016/S2665-9913(24)00190-5

Conclusion

[1] Manzi S, Raymond S, Tse K, et al. Global consensus building and prioritisation of fundamental lupus challenges: the ALPHA

project. *Lupus Sci Med*. 2019;6(1):e000342. Published 2019 Jul 19. doi:10.1136/lupus-2019-000342

[2] Mosca M. Should PROs be incorporated in the response evaluation? *Lupus Sci Med*. 2022;9. doi:10.1136/lupus-2022-la.11

[3] Tse K, Sangodkar S, Bloch L, et al. The ALPHA Project: Establishing consensus and prioritization of global community recommendations to address major challenges in lupus diagnosis, care, treatment and research. *Lupus Sci Med*. 2020;8(1). doi:10.1136/lupus-2020-000433

[4] Feldman CH, Speyer C, Ashby R, et al. Development of a set of lupus-specific, ambulatory care-sensitive, potentially preventable adverse conditions: A Delphi Consensus Study. *Arthritis Care Res*. 2019;73(1):146-157. doi:10.1002/acr.24095

[5] Cornet A, Mazzoni D, Edwards A, et al. Coping with systemic lupus erythematosus in patients' words. *Lupus Sci Med*. 2022;9(1):e000656. doi:10.1136/lupus-2022-000656

[6] Furie R. 17 Lupus treatment in the next decade: the next decade is upon us. *Lupus Sci Med*. 2021;8. doi:10.1136/lupus-2021-la.17

[7] Van Vollenhoven RV. 10 Lessons from RA: RA drug development has advanced at a more rapid pace than SLE: What can we learn from our colleagues? *Lupus Sci Med*. 2021;8. doi:10.1136/lupus-2021-la.10

[8] Wei L, Yong-gang L, Zhao Y, et al. Efficacy and safety of belimumab plus standard therapy in patients with systemic lupus erythematosus: A meta-analysis. *Clin Ther*. 2016;38(5):1134-1140. doi:10.1016/j.clinthera.2016.02.022

[9] Nalakonda G, Islam M, Chukwu V, et al. Psycho-rheumatic integration in systemic lupus erythematosus: An insight into antibodies causing neuropsychiatric changes. *Cureus*. 2018;10(8):e3091. doi:10.7759/cureus.3091

[10] Paridon B. Anifrolumab improves skin and joint disease activity in patients with systemic lupus erythematosus. *Rheumatology Advisor*. Published 2021.

https://www.rheumatologyadvisor.com/news/anifrolumab-improves-skin-rash-and-joint-activity-arthritis-in-sle-lupus-eular-2021/

[11] Ranganathan U, Merrill J, Crow M, et al. PO.6.132 A phase 2b study of afimetoran (BMS-986256) in patients with active systemic lupus erythematosus (SLE): Optimization of a lupus clinical trial design. *Lupus Sci Med*. 2022;9. doi:10.1136/lupus-2022-elm2022.153

[12] Felten R, Dervovic E, Chasset F, et al. The 2018 pipeline of targeted therapies under clinical development for systemic lupus erythematosus: A systematic review of trials. *Autoimmun Rev*. 2018;17(8):781-790. doi:10.1016/j.autrev.2018.02.011

[13] Sood A, Raji M. Cognitive impairment in elderly patients with rheumatic disease and the effect of disease-modifying anti-rheumatic drugs. *Clin Rheumatol*. 2021. doi:10.1007/s10067-020-05372-1

[14] Doria A. Targeting remission and low disease activity in SLE. *Br Med J*. 2021;8(1). doi:10.1007/s40744-023-00601-w

[15] Urowitz M. 02 Hydroxychloroquine myopathy: cardiac and skeletal muscle toxicity. *Lupus Sci Med*. 2021;9. doi:10.1136/lupus-2021-la.2

[16] Hanly J, Robertson J, Kamintsky L, et al. Functional connectivity, enhanced blood-brain barrier leakage and cognitive impairment in systemic lupus erythematosus. Presented at: American College of Rheumatology Convergence 2021; November 3-10, 2021. Abstract 0456. https://acrabstracts.org/abstract/functional-connectivity-enhanced-blood-brain-barrier-leakage-and-cognitive-impairment-in-systemic-lupus-erythematosus/

[17] Konig MF. The microbiome in autoimmune rheumatic disease. *Best Pract Res Clin Rheumatol*. 2020;34:101473. doi:10.1016/j.berh.2019.101473

[18] Ruiz-Irastorza G. Glucocorticoids for SLE: what are the current questions and issues? *Lupus Sci Med*. 2022;9. doi:10.1136/lupus-2022-la.1

[19] Venkatadri R, Sabapathy V, Dogan M, Sharma R. Targeting regulatory T cells for therapy of lupus nephritis. *Front Pharmacol.* 2021;12. doi:10.3389/fphar.2021.806612

[20] Birt JA, Hadi MA, Sargalo N, et al. Patient experiences, satisfaction, and expectations with current systemic lupus erythematosus treatment: Results of the SLE-UPDATE Survey. *Rheumatol Ther.* 2021. doi:10.1007/s40744-021-00328-6.

[21] Aringer M, Arnaud L, Furie RA, et al. Real-world treatment patterns and clinical characteristics in patients with moderate-to-severe systemic lupus erythematosus: an analysis of the SLE Prospective Observational Cohort Study (SPOCS). *Lupus Sci Med.* 2025;12(1):e001336. doi:10.1136/lupus-2024-001336

[22] Felten R, Scher F, Sibilia J, Chasset F, Arnaud L. Advances in the treatment of systemic lupus erythematosus: From back to the future, to the future and beyond. *Joint Bone Spine.* 2019;86(4):429-436. doi:10.1016/j.jbspin.2018.09.004

[23] Bakshi J, Segura B, Wincup C, Rahman A. Unmet needs in the pathogenesis and treatment of systemic lupus erythematosus. *Clin Rev Allergy Immunol.* 2018;55(3):352-367. doi:10.1007/s12016-017-8640-5

[24] Venkatadri R, Sabapathy V, Dogan M, Sharma R. Targeting regulatory T cells for therapy of lupus nephritis. *Front Pharmacol.* 2021;12. doi:10.3389/fphar.2021.806612

[25] Hirohata S, Kikuchi H. Role of serum IL-6 in neuropsychiatric systemic lupus erythematosus. *Am Coll Rheumatol.* 2021;3(1):42-49. doi:10.1002/acr2.11217

[26] Volansky R. TULIP: "Anifrolumab exhibits 'many more pluses than minuses' in lupus." *Healio.* Published April 21, 2021. Accessed April 27, 2021. https://www.healio.com/news/rheumatology/20210420/tulip-anifrolumab-exhibits-many-more-pluses-than-minuses-in-lupus

[27] Wang Y, Xie X, Zhang C, Su M, Gao S, Wang J, Lu C, Lin Q, Lin J, Matucci-Cerinic M, Furst DE, Zhang G. Rheumatoid arthritis, systemic lupus erythematosus and primary Sjögren's syndrome shared

megakaryocyte expansion in peripheral blood. *Ann Rheum Dis.* 2022;81(3). doi:10.1136/annrheumdis-2021-220066

[28] Nezhad MS, Seifalian A, Bagheri N, Yaghoubi S, Karimi MH, Adbollahpour-Alitappeh M. Chimeric antigen receptor-based therapy as a potential approach in autoimmune diseases: How close are we to the treatment? *Front Immunol.* 2020;11. doi:10.3389/fimmu.2020.603237

[29] Humphreys C. Intestinal Permeability. In: Pizzorno JE, ed. *Textbook of Natural Medicine.* 5th ed. Elsevier; 2020:166-177. doi:10.1016/B978-0-323-43044-9.00019-4

[30] Fasano A. Zonulin, regulation of tight junctions, and autoimmune diseases. *Ann N Y Acad Sci.* 2012;1258(1):25-33. doi:10.1111/j.1749-6632.2012.06538.x

[31] Felten R, Sagez F, Gavand P, Martin T, Korganow AS, Sordet C, Javier RM, Soulas-Sprauel P, Riviere M, Scher F, Poindron V, Guffroy A, Arnaud L. 10 most important contemporary challenges in the management of SLE. *Lupus Sci Med.* 2019;6. doi:10.1136/lupus-2018-000303.

[32] Tamilou F, Arnaud L, Talarico R, Scire CA, Alexander T, Amoura Z, Avcin T, Bortoluzzi A, Cervera R, Conti F, et al. Systemic lupus erythematosus: state of the art of clinical practice guidelines. *Rheumatol Musculoskelet Dis Open.* 2018;4(1). doi:10.1136/rmdopen-2018-000793

[33] Harrison M. Thinking, memory, and lupus: What we know and what we can do. *Hospital for Special Surgery.* Published 2006. Accessed April 27, 2021. https://www.hss.edu/conditions_thinking-memory-lupus.asp.

[34] Mani A, Shenavandeh S, Sepehrtaj SS, Javadpour. Memory and learning functions in patients with systemic lupus erythematosus: A neuropsychological case-control study. *Egypt Rheumatol.* 2015;37(4):S13-S17. doi:10.1016/j.ejr.2015.02.004

[35] Thomas, Donald E. (2014). The Lupus Encyclopedia: A Comprehensive Guide for Patients and Families. Baltimore, MD: A Johns Hopkins Press Health Book. https://a.co/d/8Vp2WES

[36] Panopalis, P., Julian, L., Yazdany, J., Gillis, J.Z., Trupin, L., Hersh, A., Criswell, L.A., Katz, P, & Yelin, E. (2007). Impact of memory impairment on employment status in persons with systemic lupus erythematosus. *Arthritis & Rheumatology,* 57(8), 1453-1460. DOI: 10.1002/art.23090

[37] Olah, C., Schwartz, N., Denton, C., Kardos, Z., Putterman, C., & Szekanecz, Z. (2020). Cognitive dysfunction in autoimmune rheumatic diseases. *Arthritis Research & Therapy,* 22, 78. DOI: 10.1186/s13075-020-02180-5

[38] Raghunath, S., Glikmann-Johnston, Y., Morand, E., Stout, J.C., & Hoi, A. (2021). Evaluation of the Montreal Cognitive Assessment as a screening tool for cognitive dysfunction in SLE. *Lupus Science & Medicine*; **8**:e000580. DOI: 10.1136/lupus-2021-000580

[39] Panopalis, P., Julian, L., Yazdany, J., Gillis, J.Z., Trupin, L., Hersh, A., Criswell, L.A., Katz, P, & Yelin, E. (2007). Impact of memory impairment on employment status in persons with systemic lupus erythematosus. *Arthritis & Rheumatology,* 57(8), 1453-1460. DOI: 10.1002/art.23090

[40] Butt, B., Farman, S., Khan, S., Saeed, M.A., & Ahmad, N.M. (2017). Cognitive dysfunction in patients with Systemic Lupus Erythematosus. *Pakistan Journal of Medical Sciences,* 33(1), 59-64. DOI: 10.12669/pjms.331.11947

[41] Masley, S. M. (2018). The Better Brain Solution: How to Start Now at Any Age and Prevent Insulin Resistance of the Brain, Sharpen Cognitive Function, and Avoid Memory Loss. New York: Alfred A. Knopf. https://a.co/d/9zg2Ixo

Epilogue

[1] Hahn BH, Singh RR, Wong WK, Tsao BP, Bulpitt K, Ebling FM. Treatment with a consensus peptide based on amino acid sequences in autoantibodies prevents T cell activation by autoantigens and delays

disease onset in murine lupus. *Arthritis Rheum.* 2001;44(2):432-444. doi:10.1002/1529-0131(200102)44:2<432::AID-ANR62>3.0.CO;2-S

[2] Talotta R, Atzeni F, Laska M. Therapeutic peptides for the treatment of systemic lupus erythematosus: a place in therapy. *Expert Opin Investig Drugs.* 2020;29(8):835-843. doi:10.1080/13543784.2020.1777983

[3] Spain E. Putting lupus in permanent remission. Northwestern University. Published 2013. Accessed November 10, 2021. https://news.feinberg.northwestern.edu/2013/11/lupusremission/

[4] Rosenberg A, Mangialasche F, Ngandu T, Solomon A, Kivipelto M. Multidomain interventions to prevent cognitive impairment, Alzheimer's disease, and dementia: From FINGER to World-Wide FINGERS. *J Prev Alzheimers Dis.* 2020;7(1):29-36. doi:10.14283/jpad.2019.41

[5] Aghjayan S, Bournias T, Kang C, et al. Aerobic exercise improves episodic memory in late adulthood: a systematic review and meta-analysis. *Commun Med.* 2022. doi:10.1038/s43856-022-00079-7

[6] Beidelschies M, Alejandro-Rodriguez M, Ji X, Lapin B, Hanaway P, Rothberg MB. (2019, October 2). Association of the Functional Medicine Model of Care with Patient-Reported Health-Related Quality-of-Life Outcomes. *JAMA,* 2(10):e1914017. DOI: 10.1001/jamanetworkopen.2019.14017

Want to Work Directly with Me?

I can't wait to hear your story, and begin working together! Simply go to my website https://www.functionalautoimmunity.com/ and fill out the contact form or email me directly at: julie@functionalautoimmunity.com
I offer a free 20-minute discovery call to answer any questions before we get started.

More About Me

I am a ReCODE 2.0 Certified Health Coach, a National Board Certified Health & Wellness Coach, a Functional Medicine Certified Health Coach, a Positive Intelligence® Coach, and a Certified Gluten-Free Practitioner. In addition to my coaching credentials, I am also a National Board-Certified Speech-Language Pathologist and an author. I have been fascinated with the brain since high school! I earned bachelors, masters, & doctorate degrees in speech-language pathology, with my doctoral work focusing on neurogenic brain disorders. After evaluating and treating people with dementia, stroke, and traumatic brain injury (& many other diseases) for 20 years, I developed six autoimmune diseases myself, and had significant "brain fog" or cognitive decline from lupus. My conventional doctors offered no solution for cognitive impairment, so I did my own research, and this book is the end product of that. By applying what I was learning, I ended up making diet and lifestyle changes. With these changes, my cognition cleared, and I was also able to stop taking nine medications, feeling better than I had in 10 years. My rheumatologist began to send me others with brain fog to help, and I went back to school, graduating from the Functional Medicine Coaching Academy. My business, Functional Autoimmunity, LLC, offers individual functional medicine coaching through telehealth for brain health, autoimmune disease, and, often, the overlap between the two. My approach involves developing a holistic picture of a client's health, helping them discover what they want and why, and then supporting them in making tiny, manageable changes toward the life they want. I believe small changes make a huge difference, and there is always a reason to hope.

    Not sure if coaching is right for you? Let's talk about it at no cost with a free discovery call! If you're not ready for that, email me to join my mailing list, in which you get monthly newsletters with the latest research and brain health tips: julie@functionalautoimmunity.com

Remember, you CAN do something about your brain health and your autoimmune disease. There are many things you can try, and a lot of this is under your control. NEVER give up.

www.ingramcontent.com/pod-product-compliance
Lightning Source LLC
Chambersburg PA
CBHW070610030426
42337CB00020B/3737